A VERY LETHAL INTERVENTION

T.S. KNIGHT

CONTENTS

Prologue v

Chapter 1—A Violent End 1

Chapter 2—Early Years of Savants 8

Chapter 3—We All Have Some Motivation 13

Chapter 4—The Game 20

Chapter 5—A Sinister Glitch? 28

Chapter 6—The Discovery Process 38

Chapter 7—Pandora's Box 44

Chapter 8—Revelation 52

Chapter 9—The Usual Finger Pointing 57

Chapter 10—Wait... Who Are You Again? 68

Chapter 11—Again... This Time with Plausible Deniability . 75

Chapter 12—A Moral Dilemma...or...No? 85

Chapter 13—Reflection 94

Chapter 14—The Proverbial Roller Coaster 99

Chapter 15—The "Family" Wedding 108

Chapter 16—Yin Yang 117

About the Author 127

PROLOGUE

There are definitely times when we are all faced with decisions that pose dilemmas. If we are perched on top of a fence and are quite stable and can fall to either side of our choosing, but we will definitely fall, will we choose the side where we land on pieces of broken glass or the side with shards of sharp metal. Let's add to the problem. Let's say that whichever side we choose for ourselves, we will be accompanied by many other people; most people, in fact. When we fall, only a few will fall on the other, presumably more disastrous side. Again, regardless of where we fall, we will certainly fall. How do we decide? What do we consider? How will we feel about ourselves, later?

This is the decision facing a brilliant young couple, whose only aim was to create a fun video game to educate and motivate people to make corrections, now, regarding the deleterious effects of climate change on us all. Hudson and Jamie grew up in adjacent homes in a nice, suburban neighborhood. Their parents were well educated, hard working and attentive people; albeit somewhat different in their life's goals. The two young adults had never been exposed to any sort of crime, in their lives. Their minds were pretty much unencumbered and free to focus on the "Yin" in life. There were no life-altering decisions to be made. This, of course, doesn't last forever, for any of us. The "Yang" will, at some point, rear it's ugly, but inevitable head. Plots of climate change, technology, romance, murder and especially political and corporate power highlight the essence of the story and present the readers with a dilemma, of their own… I'm a good person, but… what would I do? Also, at what point, do we realize that the usual political and corporate landscapes, of the world are often completely self-serving, ineffective, and modulated by individuals who were raised poorly and become autocratic with a high predilection for mental illness? Some of this can be explained by the study of evolutionary psychology, but not all of it. So, the question in this story is which team do we find ourselves cheering for? The answer may slowly creep up and surprise the reader and be a cause for some soul searching.

As the characters develop in this book, you will align, with them, in your own way, depending upon who you are, as an individual. As other main characters are introduced, you will start to question what you previously knew to be "right". Some sinister characters, in life, are just plain "bad", through and through … or so you thought. The ancient Chinese philosophy of Yin & Yang would dispute this belief, however and you will start to understand this better, as the story progresses. Is anyone ever completely benevolent or malevolent?

This book is a great read for young adults, techies, romantics, climate activists, philosophers, crime analysts and journalists. This broad swath reflects contemporary life, as we all know it. Timely and very compelling, it will leave you wanting to ask, as a friend, Neil, did at lunch, "… and then what happened"? .... Well … we'll see.

T. S. Knight

CHAPTER 1

A Violent End

The winds of change are blowing the tree branches forcefully, outside the office building windows. This was unusual, for late summer, but had become more noticeable to most residents, who were not self-absorbed. The remaining few hardly even notice the hand writing.

Three PetroMix company execs are having a late meeting in the board room after an all-afternoon discussion with the board on the merits and need to open a new oil drilling facility in a hotly disputed area of Alaska. It's disputed because of Native American rights to the land, proximity to a nearby town and environmental concerns. Someone in the room, briefly and half-jokingly, also brings up an issue about the disappearance of another company's CFO (chief financial officer) after the company agreed to a similar project. The person who raised the issue gets booed and laughed at, and the CEO dismisses it out of hand, saying, "Oh, ha, ha, John. Everyone settle down. Let's get serious here. We've been at this all day.

After an exhausting level of discussion, which we all know to be common when more than one person is making a decision, the board finally agrees to leave the decision to the top three executives, including the CEO, since the board was split. The meeting is concluded and the top three stay behind to hash out the decision. This smaller meeting, after hours, takes on a decidedly different flavor with the CEO becoming very heavy handed and essentially threatening the CFO and COO with dismissal from the company if they don't agree with "his baby." The CEO is quite full of himself, lacks compassion and empathy for others and is generally autocratic, all sociopathic tendencies. They both comply and feel it's not worth arguing over such a small matter as one more oil project. The CEO, who is now gloating, and the other two executives sign the agreement and then call it a day. The CFO and COO cannot wait to get away from the jackass-in-charge. One comments, "can you believe this guy?" The other just looks at him and shivers.

The CEO, still gloating but now also pissed because he had to even spend his precious time discussing the matter, proceeds down to the building's private garage, where he is picked up in his Bentley by his personal driver. After he is in the car and tells his driver to take him directly home because it has been a long day, the driver turns and shoots him in the head three times. He then drives out of the garage and proceeds to a pier at a nearby shipping port. Paying someone there through the driver's side window, he drives the car onto a small ship, parks it in a shipping container, covers it and leaves the ship. Later that night, the ship leaves port and several hours later, the ship's crane is seen picking up the container and dropping it in the ocean.

The next morning, the PetroMix CFO and COO are in the office trying to find the CEO to, once again, discuss their reluctance to open the new oil project. They can't locate him. His wife states he never returned home last night, but also notes annoyedly, that he's been doing that more often lately. She suspects an affair and seems disgusted. She does note that his not arriving at the usual time in the office the next morning is not typical for him. Now, that same morning, Hudson Harthun and Jamie Broadmoor are both getting up, having coffee and getting signed in to their online game administration site. They both are a little stunned; it happened again. A player has increased his game status dramatically and, at the same time, has probably made a ton of cash. Jamie picks up her cell and calls Hud, just as he's calling her. They finally connect. After a one word greeting of "Hey!" there is an uncomfortable silence.

Jamie starts in a very low volume, somewhat frightened tone, "Well, did you see?" "Yeah," replies Hud. "Do you think this has to do with that oil jerk in the morning news?" says Jamie.

"Do you?"

"I dunno…possibly…I mean, I guess so. It's hard to ignore the pattern, right? So, if something bad happened, that guy had it coming, didn't he? " replies Hudson. "I am getting really uncomfortable, with all this, Hud. I mean, should we be talking on cell phones, using our names and all?" says Jamie."Well, says Hud, "let's get ourselves together and meet for another coffee at that place we found last Friday morning. It's pretty public but also quiet and we can talk…say, in 45 minutes?"

"You mean Ruurlo's?" asks Jamie.

"Jeez, man. You'd make a terrible Anastasia. The Bolsheviks would find you in 30 seconds," says Hud, somewhat laughingly.

"Sorry, sorry," says Jamie. "I'm getting a little nervous."

The two meet at Ruurlo's and sit in a quiet corner booth. It's early and there are not many patrons there yet. Ruurlo's is like an old diner, with the red plastic upholstery on the seats and the chrome coat stands near each booth. Jamie and Hud are both fond of retro eateries. The waitstaff are waiters and waitresses, not "servers." Very old school. After ordering coffee and some eggs from the waitress, who hardly notices them at all, they start to discuss the matter at hand. Jamie does appear a little nervous. They have designed a game that may have gotten away from them. They're not sure. Clearly, something is happening to people "in power" who are not only not paying attention to environmental and climate issues, but who are also actually speeding up the extinction of the human race. They honestly wonder, together, if there is a direct connection to their game, or is it just a coincidence and worldwide climate activism is finally raining down on those with sociopathic personalities who are drunk with authority and profits? Either way, being climate activists, themselves, they are of two minds. On one hand, they think it's high time the climate change deniers "got theirs," but, on the other hand, neither of them was raised to take matters into their own hands in a violent manner. The two finish their conversation and eggs, pay the check and leave Ruurlo's. They notice, on their way out, that someone is sitting just two booths away from them and within earshot of their conversation. They look at the very well-dressed, middle-aged couple as they pass by. The couple gets very quiet, suddenly, and stares at them. Jamie is freaked out by this, and when outside the restaurant, says to Hud, "Do you think they heard us? I didn't even see them come in. That's weird, right?"

Hud tells her she is becoming paranoid. "What's wrong with you? They were probably having a fight and stared at us, just to break the tension. Calm down, Anastasia. They had nothing to do with us." Weeks later, the oil executive has still not been found.

On a very sunny, summer morning just outside of Denver, Colorado, Janice and Richard Harthun are scrambling around trying to have a reasonable breakfast and getting ready for work. Cherry Creek has a lot to offer the Harthuns. It is not in the city itself and has a bit more of a country feel to it. They have a nice home, nicer than their job income could offer. When Richard's parents passed on, they received a sizable inheritance and put most of it into their Denver home; the rest they invested. The fact that they decided to buy a nice home in a fashionable suburb was somewhat out of apparent character for them. These are somewhat "earthy" people, trying to live an environmentally conscious and sustainable lifestyle, composting and recycling and contributing to like causes. At home, they dress somewhat like farmers; no offense to farm families intended, but being well dressed and stylish is only for the sake of their jobs, not for life at home. Their motivations and demeanor, at home are quite different.

Richard is an editor for a local newspaper which has an online presence and is doing reasonably well, considering how hard copy news media fares these days. He dresses in a business suit, every day, but somehow manages to look like a homeless person at work, as well as at home. Janice is the head of HR for a national non-profit food bank organization. She dresses very professionally. She finds her job stressful at times, but is highly respected at work, and she uses this respect as a platform for her other interests which are environmental concerns. When home, Janice wears sweats (nice sweats) and Birkenstocks everywhere. We all know someone like that; bless their hearts.

When the Harthuns are home, they work in the garden, maintain their home and research environmental and climate issues locally and globally. They write articles in support of being proactive on this front and occasionally contribute money and even join protests in support of their main cause. They have demonstrable disdain for those who call themselves "leaders." They see them as entirely self-serving, power freaks who have learned how to exploit "believers," whom they see as poor souls not smart enough to discern between what's good for them and what's not. They also see them as people easily led astray by autocratic snake oil salesmen. Richard even occasionally writes an article for his newspaper about this. These articles are not, ever, well received by his superiors at the news office.

The Harthuns have neighbors, of course, as they are living in a nicely developed suburb. The home to their left changes owners fairly often, and Jamie and Richard ultimately have grown tired of getting to know them all very well. The neighbors to their right, however, have become reasonably close friends. The Broadmoors have a different slant on life and a decidedly different approach. Their drives and goals are quite different than the Harthuns, whom they see as friends and whom they think of as "quaint" people. Rachel Broadmoor is an HR director for a medium-sized pharmaceutical company, and her husband Alejandro, the son of parents who emigrated from Madrid, owns his own tech company, designing hardware switches for computer server farms. While the Harthun's home interior was more like a nice cabin, the Broadmoor's home looked like a set in a high-end furniture store. Somewhat minimalist and very fashionable, there was a place for everything and everything was in it's place.

Both the Harthuns and the Broadmoors are "good" people, meaning they try to do the right thing when it comes to matters affecting others directly and humanity, in general. The difference is in the level of commitment. The Harthuns will moderately go out of their way and don't much care "how it looks" to others. The Broadmoors will commit financially, but have never been the types to go to a protest or join groups supporting issues such as the environment. The Broadmoors are mainstream, religious people who follow dogma and make their occasional "appearances" at church, more to be seen by colleagues than anything else. They are very materialistic. The Harthuns are Unitarian, loathe anything dogmatic and are quite independent. They follow what they both feel is the voice of reason, with compassion and empathy. The Broadmoors are not originally from Denver. They moved there from a well-to-do suburb of Boston for Alejandro's job in the tech industry. The Harthuns, on the other hand, are both originally from Colorado, Janice from Grand Junction, a mountain town, and Richard from Colorado Springs, where his father was in the Air Force.

CHAPTER 2

Early Years of Savants

Both, the Harthuns and the Broadmoors work fairly long hours. This doesn't really bother Rachel Broadmoor; she enjoys the limelight, her corporate stature and meeting new people. It does bother Janice, however, who prefers to spend more time at home, and thus her "job" has become more of a necessity than a fulfillment of self. Richard Harthun enjoys his position at the local paper, writing and trying to sway opinion to what he thinks is reasonable. He has clear disdain, however, for the "business side" of the news media in general and sees it as an affront to truth and reason. He often puts in long hours but only if he feels motivated by a breaking story. He, otherwise, would also rather be at home with Janice, working on his or their pet projects. Alejandro, on the other hand, loves his business. He started out with nothing and worked very hard for his education. He learned his business from the ground up and built a medium-sized tech company. It is his "baby," and he is actually doing very well financially. His computer switches are state of the art, and he, himself, is adept at computer hardware and software design. His software skills are good, but his hardware design skills are extremely innovative. Alejandro puts in long hours at the office and at home just because he wants to. He was quite energetic, as many self-made entrepreneurs are.

Now, both the Harthuns and Broadmoors have other children. Hudson has two sisters, Jackie and Laurel. Jackie is three years older than Hudson and Laurel is four and a half years older. They try their best to help their little brother "clean up his act," dress "like a human" and socialize more, instead of spending so much time playing his video games. They tell him he's going to end up living in his parents' basement, playing games all day and eating pizza constantly to maintain his svelte 350 pounds. They love him, but do see him as a hopeless nerd.

Both Jackie and Laurel associate more with Jamie's sister Savanna, who is two years older, and a real fashionista. Appearances are "all important" to Savannah, who eventually goes to college for fashion design and becomes a reasonable well-known, internet-based clothing designer for her young followers. Savannah and Jamie, though, are also quite different, with Savannah not having any real bent for anything technical other than as it may relate to her clothing line. Some would describe Savannah as "superficial." Jamie and Savannah are close, though, and discuss everything from clothes (which Jamie does start to become fond of) to boys. At one point, in their early years, Savannah does occasionally mention that Hudson is a nerd and Jamie may want to branch out to meeting other boys. Jamie, though, is steadfast in her friendship with Hud and tells Savanna that he is the only one who really gets her and that they really enjoy each other's company. This is met by Savannah with the usual, "Okay, whatever. I like Hudson, don't get me wrong. I guess he's kind of a cute nerd."

When the babies of the families, Jamie and Hudson, were born, Janice and Rachel were both pregnant at about the same time, only about three months apart. Because Janice had more job flexibility, she was able to take more time off and acclimate to having her third child. Rachel, on the other hand, had a much more demanding job involving a lot of travel, and she found her second pregnancy exhausting, relying heavily on her neighbor Janice to help out with her daughter Savannah, which she was happy to do; well, not happy, but she did not find the request onerous. The ultimate result was that Janice and Rachel became closer and closer friends, in fact, both families had become quite close.

When Hudson and Jamie were born, they were both very inquisitive, "busy" little ones, almost from birth. They played with mobiles attached to their cribs for hours and spent lots of time together, mostly at the Harthuns since Rachel was often out of town on business. Hudson's older sisters often cared for them as they got older and saw Jamie as another one of "their babies," occasionally asking their mom, "Where is Jamie today?" Janice would repeat, "Oh, she's at her house today with her mom and sister. I know you love her, but you know she's not our baby, right?" Suffice it to say that Hudson and Jamie were brought up together and a lot of the Harthun's family flavor rubbed off on Jamie, who developed similar values and drives as she got older.

Hudson became an inquisitive toddler, getting into almost everything and walking at a very early age. When it came to talking, which also came early, he never stopped, his favorite word being "why?" He took after both of his parents when it came to environmental issues. He took after his father when it came to style of dress, which was not apparently important to Hudson at all (nor his father). His passions lay elsewhere. When it came to reading, Hudson was "almost gifted," to hear his father discussing this with other fathers. Hud had his face in a book, constantly, reading them over and over and always questioning. "Why does the little boy's tiger friend sometimes appear to be a stuffed animal and sometimes appear to be alive and, by the way, why did the tiger speak? Do all tigers speak, occasionally? By the way, how is it that the tiger seems to have an elaborate answer for everything? Also, by the way…" You get the picture. Hudson's accelerated level of cognition, if you will, was not only noticed by his parents. His teachers noticed it also, sometimes finding it exhausting, but always finding it challenging. Several parent teacher conferences ensued over the years, prompting a recommendation for a school counselor evaluation for ADD, which followed, but the conclusion was that he did not carry this diagnosis and was just very bright.

Hudson's interaction with other children was less than optimal. He socialized but seemed bored quickly and went back to what he was previously doing. He did seem to want to play games, especially games with a puzzle format. He wanted to play them constantly, making the other children bored. Hud also didn't care about having the latest dinosaur T-shirt or superhero cap. He liked sports at recess in grade school, but as he got older, he strayed away from them, seeing them as simple in design and boring. He associated them with his most boring friends. While he was physically adept, the only sports he enjoyed were games like soccer. He saw them as more strategic than physical. Games like basketball put him to sleep, owing to their rapidly repetitive nature.

As Hudson got older, his attention and clearly his passion, centered on math in school and video games at home, seeing them both as "something to be solved." His capabilities and level of reasoning were noticed by teachers from an early age. One of them jokingly commented to his parents that someday he would probably work for the CIA or something. Hudson did date, occasionally, starting in high school. Girls noticed his sense of humor and how open and helpful he was. It was likely that some of his girlfriends saw him as a "fixer upper" that would likely be a "great home" in the future. He saw them as "great fun," but "hey, let's be serious," when asked if he had his eye on anyone special. This was in contrast to Jamie, who, as she got older, was a strikingly beautiful girl outside with a techy nerd inside. She had long wavy blonde hair, very fair skin and bright green eyes. Everything about Jamie was in proportion, you know, the kind of girl that the other girls hate. She dated quite often, also, in high school, but she had no special person she thought of as "the one." When other guys would see her walking around with Hudson, they would comment to each other, "What the heck is she doing with that guy?"

CHAPTER 3
We All Have Some Motivation

Climate changes and the activism surrounding them have become a strik-ing push-pull issue in the world, pitting power and profit-motivated, authoritarian figures against the masses of earth's populations. Younger people, in particular, but not exclusively, see the "older generation" as not paying nearly enough attention to their future existence. These young activists have grown in number and are spurred on by leaders of movements supporting conservation and an end to fossil fuel usage. Groups supporting green energy, from solar to wind and nuclear fission to fusion and hydro, are popping up almost weekly now. Their voices are growing louder and social media is paying much more attention to them (albeit encouraged by a profit motive). Every time the media reports on wild fires, hurricanes, floods and tornadoes or on animal displacement from their normal habitat and near extinction, the activists' voices grow stronger. Nothing encourages them, however, more than the reporting of oppressive climate deniers who publicly take a stance, clearly without regard for the health of the planet, for profit and power, cloaked in a green cape of lies, which no one is fooled by anymore and which reflects the sociopathic tendencies of those who have lost touch with their humanity (or perhaps never had any). Such reporting clearly pours gas on the already raging fires of activism, the trajectory of which becomes more and more obvious to many. It's not surprising, then, that a rare violent event occurs, as a result. Consider the following:

A recent news report is focused upon the leakage of a massive amount of crude oil from a tanker ship off the coast of a Scandinavian country. The coastal communities are severely damaged and the local fishing industry takes a huge hit. As expected, a full barrage of finger pointing and lies and promises to do better follows. There are protests and short films and a follow-up responsive barrage of finger pointing. Something is different this time, though. The shipping company's owner, the CEO of the oil company and a little-known government official of the Commerce Department are said to have resigned. This, at least, is the initial news report. Little by little, weeks to months later, there are reports that one or the other of these bad actors cannot be found or are thought to have moved out of country. It isn't long before all three are never heard from again. This is months after the initial report and most paid no particular attention to this "white noise" news.

Borgesa Torgen, CEO of ScandOil, is a strong Scandinavian woman who as a child learned about the oil business from her father, who was a driven personality without a sense of compassion or understanding, even for his own daughter. He treated her like an employee after her mother died of cancer, taking her to the office often, more for seeking child care from his executive assistants than anything else. Borgesa was very young and impressionable at the time. She watched her father—how he spoke to other people and what he would say to someone else, quietly, after the other person left the conversation. He clearly had no respect for anyone. As Borgesa got older and began to firmly develop her personality, she would question her father about his interactions. He would sit her down and explain how "this was just business," not a social interaction. This was how business was done. He explained that if she wanted to get along in business, she had to be tough and uncompromising. Other people in business did not matter. It was a "you or them" scenario. He also made it clear that he would not tolerate her thinking any other way because it would not give her a "good" life. Borgesa listened…and learned.

After the oil tanker incident, Borgesa is in her office working late with her personal assistant Inga, a younger woman who has been able to survive her employ for over five years. She is not treated well by her "boss" but stays because she is well paid and has a child with her husband who had been injured on an oil rig and is now permanently disabled. She was their sole support. Other women in the company would occasionally ask Inga why she tolerated the CEO's abuse. The reply was, of course, very understandable, but sad.

Borgesa has had a long day and is cursing a blue streak. She is outraged that her executive team has been finding fault and company culpability relating to the oil spill. She is at a level of rage not seen by Inga before. After the oil spill news broke, however, Inga was expecting this. She wonders how she can tolerate the onslaught that is about to befall her from Borgesa. This night, however, is too much. Borgesa is dumping all her outrage on Inga for not stepping up to support her in front of the other executives. The verbal abuse is too much… way too much. There is a point, for most people, where they simply cannot tolerate one more second. Inga, who is normally a very nice, compassionate, well-mannered, friendly, young woman asks Borgesa—suggests, in fact—that she have a drink to calm her nerves, so that she can think better. Borgesa agrees and sits behind her huge desk in her huge leather chair to relax. The office was on the top floor; a corner office. It was impeccably decorated. Borgesa even had a multi-million dollar Van Gogh hanging on the wall. Because of this, her office was very secure with a heavy door. It was fairly soundproof. Inga goes over to the credenza behind Borgesa's chair and puts out two glasses with ice. She picks up the carafe of Scotch and suddenly, while shaking, stares at it in her hand.

"Come on, Inga. Can't you even do this right today?" says Borgesa.

Inga replies, "Sorry," and turns and brings the heavy carafe down on Borgesa's head with such great force and anger that she falls from the chair and strikes her head again against a stone sculpture behind and to the side of her desk. "Dead…she's dead," thinks Inga, staring into space for a bit. With an out of character steely demeanor, she thought "I believe I can get paid, for this and who's going to miss her anyway. I'll just pour some whiskey down her throat and then I'll post the whole thing. It's done. It's done." Inga is a player.

The next morning, Borgesa is found and after an investigation by the authorities and testimony from other employees (none of whom is "friendly" with her), the incident is surprisingly declared an alcohol-related black out with subsequent head injury. Inga, being "distraught," leaves the company. Shortly thereafter, she comes into a lot of money, but no one knows…Inga is a player. Is the investigation quickly wrapped up…because someone in the police department is also a player?

As time passed, Hudson and Jamie grew closer, owing to their Mensa level intellects and very common interests of deductive reasoning, problem solving and computer technology. This was all matched by their keen interest in the environment and climate change, owing to Hudson's parents. They breezed through the high school years, seemingly unaffected by the usual high school traumas that affect most of us. Hudson wasn't bullied because he was a video gamer and there were lots of those guys, at his school, including some of the high school sports players, who thought he was cool. No one messed with him because of that. Hudson was accepted to community college in Denver. He could have been accepted pretty much anywhere he put his mind to; however, he didn't really want to leave Denver because he saw his parents aging and he enjoyed their company and their influence. He also did not wish to "break the bank" by going to some ivy-league university, where he would later be saddled with huge debt. He saw the higher educational system as a money sink, where the top administration people made millions at the expense of students, who were then paying off their debts for decades. Hud never felt the need to have a big name college or university behind him. It just wasn't him…and he also was very confident that he could get the higher education he wanted, as long as he was exposed to the right resources.

Jamie did not have to deal with the usual high school girl traumas, either. Not wearing the right clothes was not an issue for her either. Her sister Savannah, the fashionista, saw to it that Jamie set the standard for dress, thus Jamie was very popular. Jamie got accepted to the University of Denver, where her older sister Savannah went. It was a more "fashionable" school, which Jamie was not opposed to and was a "money" school with the best connections and resources.

Hudson and Jamie also did not want to leave each other, though were very reluctant to come out and admit this to each other. That would come later. They were driven by each other's intellect and common interests, but also felt, deep inside, that they were true kindred spirits and fed off each other's energy, especially during times of need.

During their undergraduate years, they both excelled at math and computer science. Some of their contributions to their schools' artificial intelligence programs were unmatched, even by alumni engineers. Their names and papers became quite well known and respected. Hudson graduated with a BS degree in computer science, with the highest honors. Jamie did likewise, albeit with a minor in fashion, owing to her older sister, who clearly had rubbed off on her. She also had a second minor, biophysics, which she found very fascinating and saw as the future of artificial intelligence, which many others just blew off as screen play for the movies.

Subsequent to undergraduate school, Hudson went to work for a small start-up video gaming company in Boulder, Colorado. His creative genius established the small company quickly, with production and sales of several games that found almost unimaginable success, very quickly. Hudson was bored and tiring of it after only a year. Jamie went to grad school, also in Boulder, at the U of Colorado, where she received her Master's degree in Ai (artificial intelligence). She was noted by many fellow students and professors as functioning "at a savant level." When asked, early on in her post-grad education, to present her paper on bio-electric computer chips design and their place in the future of Ai, she stepped to the podium. Some snickers, from both fellow male and female students could be heard, along with a comment between two young men, "Oh, my God! Who let this bimbo in here? This should be a hilarious waste of time." Jamie was a statuesque, extremely fashionably well-dressed, beautiful young woman. Instantly, upon opening her mouth, however, the lecture hall fell silent. Even her opening sentence was a stunner.

Now a quiet comment, between the same two young men, took a decidedly different tone. "Oh…my… Godddd! Who iiiiiissss this woman?!" Jamie's voice and her sultry manner of speaking stupefied the entire hall, including some of her professors, who were briefly speechless. The content of her presentation and her concepts were so unique and so well-supported, that a pin drop was again heard, in the back of the lecture hall. She was never again taken as the "model" who somehow made it into grad school because her father bought a new building for the school, or something else equally presumptuous and insulting. Little did they suspect that there was a sweet-hearted, compassionate, empathic young activist …. Gorilla, living inside this presenter.

CHAPTER 4

The Game

As Hudson is starting to feel his employment at the video game company is becoming mind-numbing, Jamie is graduating from her master's degree program again, of course, with the highest of honors. She is offered a position, in a PhD program, but declines it. She is also then offered a teaching position, which she also declines. Both Jamie and Hudson are starting to feel their creative natures are calling for them to become involved in something bigger.

"What if we came up with a game that would educate and stimulate the players to become involved in responsible Climate management? Nothing else seems to work. All the activism we've ever seen comes to naught. It's so frustrating!" exclaims Hudson. "We could base the game system on Evolutionary Psychology and your training and ideas on Ai could drive the rewards part of the game and get more people involved. What do you think?"

"Oh, I think I'd love that, Hud. It would give me a chance to test out some of my novel ideas on the future of Ai design in a virtual world. I could see how effective it was and, even better, how well players accept the outcomes. Oh! I feel my creative juices flowing already!" exclaims Jamie, who then questions Hudson about when he would like to start this and whether he would have time, given his current employment.

"Yeah, not to worry about that, at all. I'm so done with designing these cheap mind-numbing games for kids. I'm really done. Let's do this. I'm excited about it, too."

The two decide to put their ideas quickly on paper and start work in about a week with a brainstorming session over breakfast, at Ruurlo's, which is becoming a favorite hangout for them.

The following week, they meet again, each with several pages of notes. Jamie is delighted by the overall game concept that Hudson has come up with and he is equally stunned by Jamie's creative ideas regarding the Ai design and how it may be used to educate the players, without their even knowing it, and motivate them to seek more education and information by rewarding them with higher and higher status among the other players. They decide to keep the game "proprietary," i.e., not getting support or funding from outside game companies. They want to keep control, for their own satisfaction and possibly some profit. They approach their families for financial assistance. Hudson's parents contribute what they can and his two older sisters actually put in "a few bucks." Jamie's parents, particularly her father, are very enthusiastic and fund their start-up almost entirely, the remainder coming from an online group funding platform. Jamie and Hudson are thrilled and decide to go home, refine their plans, a bit, and meet again in 72 hours, at Ruurlo's, where they are now becoming recognized as regulars now. Even the waitress knows their names and finds them the "cutest, but weirdest" couple in conversation with other waitstaff. "Yeah, Beth, this young woman is such a brickhouse and the guy she's with ... the guy with the 5 o'clock shadow and the mussy hair is really intriguing.

You gotta meet these two, sometime. They're the cutest couple I've seen in a long time, considering the way kids dress these days … you know, the guys with their pants half way down their ass and the girls with nose rings and more tattoos than a sailor, back in the day."

Their next meeting is decidedly more organized. They both have their notebook computers with them, and their ideas are much more refined. They decide they should speak more quietly, especially if anyone is within ear shot. Hudson starts. He describes a virtual environment, with many players globally, with the ultimate goal of not only limiting the climate change trajectory, but reversing it. To achieve this, players would be subtly educated by attending and then becoming involved in conferences, protests, marches and letter writing campaigns. They would learn about the main causes of climate change and would see interspersed very brief videos about the results of human failure to care for the planet. These would be 5 to 10- second clips, selected by Hudson for their impact on a player's emotions, particularly fear. The clips would then be followed by several possible "actions," the most effective of which would give the player points. Significant numbers of points would then translate into a higher gamer status, of which there would be many. Each higher status icon would become harder and harder to achieve, the highest status being almost impossible. Therefore, players could collaborate on achieving higher status and, thus, share the points.

The competitiveness of Hudson's game design is striking. He feels, though, that insulation of players from each other would be necessary to prevent point theft. He has the idea of using a global platform, already in existence, as an insulator. He decides on a concept related to blockchain technology, used for management of cryptocurrency trading. Each player would have a "decentralized smart contract" ensuring the reliability of their successes and rewards, sort of a WEB3 platform. The players would enter the game with a subscription fee of about $20 annually (which Jamie and Hud would use for game maintenance and, perhaps, some profit). Each player would then create a public user name of their choice, and the game platform would issue a highly encrypted password, known only to the game administrators and only temporarily saved on company servers. Hudson would draw his ideas from real-world events to enhance reality. He would be careful to change the names of real-world autocrats and climate deniers. This is a game, after all.

Jamie listens carefully to Hudson's concept, her brain now churning with ideas. When Hudson is finished, she starts to lay out her ideas on how her Ai design could enhance each of his game steps and goals. Hudson's jaw is somewhere between the table and the floor of Ruurlo's as he listens to Jamie. "Holy crap? Can you really do that? How did you come up with that? Will it really work? To quote from *Jaws*, 'We're gonna need a bigger... computer.'"

Jamie laughs and, in her way, flicks her hair and says, "Yes, dahhlinggg. It's nothing, really." They both laugh so loud, the waitress comes over and chuckling, tells them she'll have to throw them out if any more patrons complain. She gestures around the room. There is, of course, no one else in the restaurant at this time.

Jamie goes on to describe her ideas. "The fundamental hardware design would be a biological computer chip. Actually, it would be a hybrid of a silicon semiconductor and human stem cell generated neurons. It would be much more compact, much better at storing large amounts of data and would generate a lot less heat than standard computer chips, which are very limited now because of their size requirements and the heat they give off. These chips are much better at decision analysis, also. This was all in another paper I was going to present before they literally threw a Masters, at me, but I literally ran out of time and figured I'd use my research, myself, at a later date. My point is, yes, this works and we could use it to store info for game players, make decisions about how to reward players for their climate accomplishments, enhance players' abilities to collaborate, enhance players' climate education and subliminally motivate them to become more and more motivated to do better with their next challenge. Your job could include the software UI (user interface) so that the players would have fun and get excited while they learn."

Hudson is overwhelmed, but remains calm. He asks Jamie some technical questions about how she can control the processing direction of the biochips. Jamie states she feels that it will be a bit of trial and error in design and results, but believes she can control it and achieve a previously unknown level of game reality.

"Well, we're gonna need a lab, for this. Yeah, and a business plan. Let's get started on this today," states Hudson.

The two then revisit the funding they got from their families and decide that they don't have enough resources for both the computer lab and servers and also the fancy bio-semiconductor lab Jamie will need. Then there is the question of hiring some technical staff for Jamie. They both decide that a minimally detailed game description and plan could be put out for some crowd funding. They decide to start out asking for the sky and requesting a million; they are shocked to find out that their plan description brings in close to 3 million and both start to believe the current public sentiment regarding the climate, coupled with major distrust and dislike for political leaders everywhere, has created a huge market for their game.

They decide to name their company Harthun-Broadmoor Video Entertainment. Their parents are delighted and because they are grassroots contributors, they ask if there's anything they can help with. Jamie's older sister believes she could play a role in costume design, especially for the female game characters. Jamie's father, having a fairly substantial background in computer switch design, is very enthusiastic. Hudson's parents would like to help create the user manuals for the game, and his mother would like to help manage any staff, along with Jamie's mother, since they get along very well and are both HR people. Jamie and Hudson have huge reservations about involving family, but since they see these roles as largely peripheral, they agree, initially, anyway.

Soon, the labs are created in a small space just outside downtown Denver. Jamie and Hudson are thrilled and energetic and get to work right away. The two families are anxious to get involved in this "fun" venture, but after offering some initial ideas, they are told it's way too early in the game development to utilize them. They are gently told that their ideas will be applied at the right time. "Now, go sit over there until we call you," Hud and Jamie say in a very diplomatic and "sweet" way, as much as you can do with family.

The name of the game will be CORPUS-A (collaboration of responsible people for us-all). Jamie and Hud also add v1.0 for version one. Their excitement now exceeds their energy levels, and they need to take frequent breaks, making more money, of course, for Ruurlo's, which has become a home away from home. They have become good friends with the owner, Gerrit (Garrett) Ruurlo, whom Jamie likes to call Gary, which Hudson thinks is disrespectful, but Gerrit doesn't mind.

Over time, the pieces start coming together. Hudson designs a bright, realistic user interface, which Jamie is able to modify, as the game progresses, with her Ai. Games generally start out with a goal and step-by-step incremental problems to be solved. The problems and the successful solutions are all pre-programmed, as are the possible selection of characters. So, if a goal is to get from A to Z, the first step may be to go from A to B. A game may have pre-established methods to accomplish that and a player would have to select the right one and would then be rewarded. If the player selects the wrong one, there may be a penalty, such as a decreased point total or a status demotion, etc. Everything that is needed is generally contained in the software of the game. There really is no flexibility.

CORPUS-A is quite different, however. The characters can be as realistic as a player wishes, right down to importing their own pictures into the game. Some select this, but many do not. The game suggests not doing this in order to maintain anonymity of financial reward. The game's ultimate goal is to see a decline in global temperatures and greenhouse gas emissions. The incremental goals are to address threats to the ultimate goal, one at a time. These incremental goals are to address fossil fuel threats, such as from power generators (coal, natural gas, etc.). The goals are initially pre-programmed, as are the solutions. Jamie has designed her Ai, however, to "learn" about a player's "bent" and capabilities. The program will then suggest lesser or more significant incremental goals to that player and suggest more or less aggressive solutions. The Ai may also then subliminally motivate players to take a certain path. The game then becomes more of a "living" game, where there is ultimate flexibility. The game actually will use the internet to scan media and business and governmental sites for new or next level incremental goals and add them to the game, enhancing realistic scenarios in a "virtual" environment.

Version 1.0 becomes a big success. Jamie and Hudson realize their personal goals of developing a real-world type video game which is not only fun to play, but highly educational and which strikes a major chord with Gen X and Z players, initially. Within a year of release, their company breaks even financially, and shortly thereafter, players' blogs show they are clamoring for a new version. The player base is now just over 7 million players globally. Despite maintaining the low game subscription fee of $20, Hudson estimates the game fund is about $130 million, and Jamie and Hudson's company's share is about $10 million. They now have a full staff of basic programmers, and their families are all involved in the business in very specific but minor ways, such as graphic design, computer lab maintenance, HR departments, etc. Success resonates in a big way with Jamie and Hudson, and they are more energized than ever to start on version 2.0. Back to Ruurlo's.

CHAPTER 5

A Sinister Glitch?

The push for strong global climate improvement measures has gotten stronger and stronger. This, coupled with Gen X and Gen Z's discontent and anger about climate control lip-service from those in high corporate echelons and government agencies, is creating a previously unseen level of global activism. This appears to be everywhere, but is focused most loudly and aggressively in the countries felt to be most responsible and least responsive. The days of the lip-service tranquilizers seem to be coming to an end.

On a sunny winter's morning, a North American government official, with his pockets being well-lined by the coal producers, blocks a piece of green legislation, which is considered to be among the first to deal almost lethal blows to the fossil fuel industry. This certainly delights the coal industry as well as the government official. It is also seen as a reason for the fossil fuel companies and government and media PR influencers to celebrate their ill-gotten gains, once again. There seems to be no end to the corruption.

Later that same day, the government official is meeting, with other public and private officials, for a celebratory dinner. Public scrutiny, now being what it is, deters such a questionable get-together from being held in a restaurant or even private club, and a prominent coal mining executive decides to have the get- together at his home, which is a beautiful country estate surrounded by a tall stone wall with a manned security gate. He has excellent security for the event, the guest list of which includes government influencers, a few other government officials, other coal mining company executives and families. The estate grounds are beautiful, green, rolling hills. It's quite dark outside and looking inside the home, it appears to be glowing, for the affair and is quite castle-like. The lavish affair commences and all are having a wonderful time. Inside the front foyer and in the dining area are several plain-clothes security agents. During the after dinner get-togethers, the men move to the billiard room, where it's cigars and brandy. During the conversations, a security guard whispers in the government official's ear that he is needed on the phone for an urgent matter. The official excuses himself, saying, "Gentlemen, I'll be right back." He chuckles. "There is no end to the need for my services." Yeah, another self-serving, autocratic blow hard. Once down the long hallway, accompanied by the guard, he is directed into a room to use one of the "secure" house phones. Once inside, a plastic bag is quickly put over his head, he is struck with a hammer and dragged away past a service area and out back into a waiting black van marked: "Discretionary Extermination, Inc."

The guard returns to the billiard room and tells the others that it looks like the previous gentleman's call is going to take a while. After a period of time passes, the homeowner and coal company executive decides to find out what's keeping his comrade. As he leaves the room, the guard states, "Oh, I'll accompany you, sir." This is met with a "Yes, that's fine. Thank you. Now which room did he take the call in?" As soon as the homeowner is in the fateful room, a plastic bag is put over his head, and he is struck over the head with a hammer repeatedly. He is dragged to the black van, which has now returned. The van takes off again to the same mortuary, where the two bodies are scheduled for cremation later that night.

It takes very little time before the two missing men are causing a stir among the guests. The night's festivities seem to have been cut short when the police are called to help find the two gentlemen. After a strong "dog and pony show" at the house, resembling an amateur murder mystery, the other party attendees are advised that they are not to leave town and that they will be contacted as soon as there is any information at all. After the inevitable and useless protests, the two involved families are advised that their house phones will be monitored for the next few days in case anyone calls with demands.

As the police leave and are outside, in the night air, one officer leans over to the senior officer and whispers, "You really sold that. Are you really going to do any of this?"

"Nope," replies the senior officer.

The families of all concerned are more than distraught and demanding of everyone, and, therefore, the whole affair manages to leak to the press in a couple of days. The fact that a major coal producing executive and, now especially, a government official have both gone missing is front page news. Conspiracy theories abound. Were they cheating on their wives and reneging on promises made to concubines? Were they gay and ran off with each other, rather than be found out and dragged through the media? Was a high level fired employee involved? The barrage of modern news media reports have created all sorts of conspiracy theories and possible explanations which become memes, in the minds of the ravenous public, around the world.

There are reports of sightings, almost like Elvis, and a large cash reward is offered for information leading to their whereabouts. The usual investigative theater plays out, but the police are fairly close-mouthed, simply stating that there has been no new information except that which, as usual, comes up empty. Weeks drag into months and no information of substance turns up. As is the usual approach, less and less is said about the story, with the media preferring new, fresh stories—the alarming rise of the price of coffee and gasoline and government pundits claiming that so much is being done to curtail the use of nuclear energy. More attention is paid to government sex scandals, than to climate events that could effect all our lives.

Across the globe, however, other, not dissimilar stories, are playing out, albeit at a lesser level of public scrutiny. The suspicious death of a young green energy activist in China has caught the attention of other activists in Asia. The family of a small natural gas equipment manufacturer in Europe is getting credible death threats. A Canadian shale producer executive is said to have gone missing. None of these events has drawn much global media attention, however. It seems to have become media filler, background white noise that goes in one ear and out the other; very little sticks in the middle. That is, until about six months after the coal incident involving the two missing fat cats. Now, there emerges a story about an oil refinery executive group, a think tank for the petroleum industry, so to speak.

The PetroGo refinery is one of the largest in the world and is on the subcontinent. It is a major contributor to the Oil Refinery Companies of Asia (ORCA) group. ORCA's primary role is that of a "spin" PR organization, although it cloaks that in the greenwash of the current jargon of climate activism, hoping it will be perceived as a fossil fuel organization that "is trying hard to go green." Nothing, of course, could be further from the truth. ORCA is responsible for public "damage control" when an oil producer or anyone, actually, in the chain of oil extraction, purification and transport is accused or found to be drawing public fire for negative effects on the climate. They are not a group on the public radar but work quietly through mainstream media, social media and grassroots groups, who spend time in the communities, the universities and the film industry creating a stream of public consciousness that is favorable to ORCA's member companies. They have been very effective and incognito, until now.

The PetroGo refinery is now testing the ultimate effectiveness of ORCA. In the spring, several "cracking towers" catch fire, and feeder lines from huge storage tanks also rupture, creating the largest refinery fire in history. This fire is mismanaged from the start, and it takes over a week to finally control the burn. A large amount of toxic smoke spills into surrounding communities, overwhelming the local healthcare facilities. Many fatalities occurr, both at the refinery during the fire and in the communities. The ORCA response is dramatic, so much so that it appears to be quite transparent. They are not an anonymous organization anymore. The more the mainstream and social media investigate them, the more is learned about how they have fed the public misinformation for many years. Member companies are "outed" as well as the staff at ORCA.

To say this does not sit well with local communities, as well as with worldwide activist groups, is one of the grossest understatements in recent history. Almost the entire executive and support staff receive death threats. Several times, activist groups try to raid the local building that houses ORCA. These attempts are put down by local police and, in one case, by the military. The "heat," so to speak, does subside after a few weeks; however, this is not the end. Over the course of several days, the entire executive staff of ORCA goes missing. One week later, they are videoed in an undisclosed location. The video is released both on social and mainstream media, showing the executives, each detailing their role, in whitewashing negative climate events over the years. They are forced to apologize to the world. Shortly, the video shows each executive being hung. The video is quickly pulled from social media, but not before activists all over the world, witnessed a pinnacle event in climate accountability.

The ORCA event is stunning. The response from the fossil fuel industry and many nations' governments is equally stunning, calling for immediate apprehension of the perpetrators, followed by the usual greenwashing of the industries involved. Despite the highest and most forceful levels of investigation, bluster and chest pounding, nothing significant turns up and in the next two months, this, too, stops appearing in the news. Not everyone lets this slide completely into the abyss, though. Conspiracy theorists on social media start putting events of this nature together, suggesting an overriding attempt to destroy all the jobs of the hardworking men and women in the fossil fuel industry who are "just trying to feed their families." Fingers are pointed at many groups, from government officials who are "just too green," to political activists trying to get "their side" elected, to climate scientists "who don't know what they are talking about." The populist movement towards anti-science, stupidity and all sorts of fascist characteristics plods forward.

All of this is just more of the usual human tendency to engage their fearful caveman genetics and circle the wagons for a fight, although not certain who the enemy really is. This, of course, is paid no heed by those with an actual intellect, except for one genuine investigative journalist, working for the open and truthful magazine, *Thoughtful* with worldwide distribution. Sara Meclin is, for whatever reason, breezing through social media content on ORCA and is in serious need of a "good lead" to follow up on. Her recent work has been less than exemplary and she feels her position at the magazine is in jeopardy. Giving it a lot of thought, she figures this may be her "shot," and she delves into the concept of some type of a secretive, organized effort to make necessary corrections for the good of the climate. She notices that despite some recent, albeit lesser, events like ORCA, "full and comprehensive" investigative efforts have never amounted to anything. She finds this lack of real accountability odd, that global national police forces and government agencies, such as the CIA, FBI and NSA in the United States, other national police forces and even Interpol, have come up empty handed. These agencies have not even come up with the usual, self-serving excuses and promises to do better next time. Was this some type of organized global crime network? Why? What was the motivation? Climate? No... no money in that. Shake downs? Globally? No. What organization would have the clout to carry that off on a planetary scale? Sara keeps exercising her brain, trying to come up with a common thread, to no avail. She plans on this being her big shot, her big story, and nothing will get in her way... nothing.

Over the holidays, Sara goes to spend them with her brother's family and her niece and nephews. She loves spending time with the kids. Her niece is growing up now and has become very interested in music and dance. She is taking lessons in both. Her two nephews are ten and twelve now and clearly little video game addicts. While the younger one is playing games with dungeons and monsters, the older one is playing a game called CORPUS-A.

Sara watches and tries to learn the game quickly, but it is rather involved and she finds it a little boring, initially. "Why are you so interested in the climate, Sawyer? I mean, that's wonderful...but wouldn't you rather be playing monster games, with your brother.... or, god forbid, your little sister?"

"No, not really," says Sawyer. "My friend got me into this and his dad plays, too!" Sawyer then goes on to elucidate all that he has learned about climate change.

Sara is very impressed, saying, "Wow! You sure know all about the climate and how we're ruining it. Your school must be really big into that!"

Sawyer's dad comes into the room saying, "Nope. I don't think they talk about that at all, in his school. He has learned all this from the game. Cool, huh? A game he loves to play and it's so educational."

Sara says, "Sawyer, I'd love to try to play this with you again later. Maybe you can show me how."

They all adjourn for dinner, after which Sawyer grabs his aunt and brings her into the den. "How much do you know about how to fix the environment we are ruining?" asks Sawyer. "What?" asks Sara, not expecting to hear such questions and concerns from a twelve-year-old. "Not that much, I have to admit," she responds.

They start to play and Sawyer teaches his aunt how to hunt for opportunities to fix the climate and how to select possible solutions and get points if she's right. She is paying very close attention now. Her sister-in-law comes into the den, after about an hour and a half, saying, "My god, Sara. You still playing this boring game? Sawyer, you have to let your aunty come talk with the grown-ups now."

Sara and her sister-in-law leave the room with Sara commenting on the game. The adults sit down with some wine and cheese to watch the evening news, which has the usual smattering of environmental/climate sound bites before it switches to the price of gasoline and electricity.

"Why don't they just go solar and be done with it?' says Sara's brother.

"Oh, that's because they don't have enough people trained in solar installation and technology to meet the need. That coupled with the fact that those workers in the fossil fuel industries like coal and petroleum don't really want to go for re-training. This, despite the fact that fossil fuel jobs can be dangerous in the short run and the long run, with diseases like black lung, for example. Then there's the fact that they make more money in the fossil fuel jobs, but green industries are rapidly catching up."

Sara's brother and sister-in-law are bewildered and her brother looks at her and laughs, saying, "Jeez, now you sound like Sawyer." They all chuckle and Sara is also bewildered, wondering where she got all that information. Did she get all that from the news, or… was it that game? She thinks more about it and realizes the steps she was taking to play the game didn't really have that much information in it… or, did it? "Why is that information so fresh in my mind as if I had taken a course or something?" she wonders. She decides to research the game a bit more, prior to arriving at a conclusion. Prior to the kids hitting bed for the night, Sara asks her brother where she can buy a copy of the game.

"Oh, it's a subscription, and well worth it," he replies. "It's only twenty bucks a year and here's the good part. Sawyer has only had the game for about four months, and he and his friends have collaborated and already scored about a hundred bucks in rewards. Can you believe it? Between the three of them… a hundred bucks? It covers all their annual subscription fees and when they want to cash out, the money can go back to my credit card or I can use it in my crypto account to buy more 'coin.'" He laughs. "Awesome!"

"What?" says Sara. "It pays children money?"

Her brother explains that the kids don't really care much about the money, but they're very "status" oriented and they try to "best" each other all the time.

"Huh, how much could an adult make playing this game?" asks Sara.

"Oh, I don't know. Listen, this is a game. You're not going to get rich here, right? It's a game," replies her brother. They all decide to retire for the night.

The next morning, Sara's memory of her new climate knowledge is still quite fresh and she is motivated to subscribe on her cell and get started researching this game, thinking this could be another good lead for a fresh article for her magazine, not yet realizing how this game and current world news events may be connected. After the holidays, when she is back at home and in her office, she pitches her idea for an article on this educational and "profitable" video game for kids and adults. She explains her experience. Her editor looks at her with his head at 45 degrees and asks what happened to her article on the ORCA hangings and the fossil fuel industry. She is about to speak when she stops in a peculiar manner.

"You okay?" asks her editor.

"Yeah… I think so. You know, I think I need to just research one more angle on that a bit, and I'll get right back to you."

"Okay, but Sara… I don't have to tell you. I can't give you much time to mull this over, right?"

"Yeah, I know… Got it, got it," replies Sara.

CHAPTER 6

The Discovery Process

Later that same evening, Sara is home, giving the game more thought. Her initial thought line is that the game offers climate problems to be solved and then offers suggestions for the player to try to implement. Also, players may offer or create their own solutions and, either way, the game will create a probability for success and a reward and player status change, if the solution proves effective. "Solutions are proved effective or are not based upon some sort of internal game algorithm," she thinks. How else could a video game judge a solution to a problem that has just occurred or has occurred since the publishing of the latest version of the game? That is, how could it internally judge the solution to a real-world event, which has only just happened? "It simply couldn't, right?" thinks Sara. She gives this another hour's thought; decides she is wrong—there couldn't be a connection between this virtual game and real events. "My boss will fire me, for sure, if I even raise this idea. This is crazy!" she thinks. "Oh, brother... it's time for bed."

The next day, she still cannot get the idea out of her head and decides she will pursue the crazy idea on her own and start writing something else for the magazine, hoping it will be enough to save her job. Arriving at her office, she grabs a coffee and, at her desk, she is visited by an office friend, another journalist who was a bit younger and fairly new with the company. Despite being a newbie, he seems very well off, by virtue of the way he is dressed and the car he drives. "Must come from money," Sara thinks.

"Neil, let me ask you something." She goes on to describe her theory of a connection between a video game and real climate-related, very sinister events, thinking that because Neil is younger, he will be more "hip," and probably play these games all the time.

Neil listens carefully and then tells Sara this couldn't be true. "No games have that kind of real-world reach, Sara. There isn't the computer processing power to do anything remotely close to that. That's like saying, if you really, really believe in a super-hero, they will somehow manifest themselves in real life,"—he chuckled quietly—"…or do you believe that?"

"No, Neil. Don't be stupid. I don't believe in super heroes… or ghosts, or Santa. Jeez. Give me a break. I just thought—"

Neil stops her right there. "I'll tell you what I think. I think you shouldn't mention this craziness to anyone, especially the boss, or you'll be cleaning out your desk in about 10 minutes. I'm serious, really. We're good friends and my advice is to stop playing video games with your brother's kids."

This discussion makes perfect sense and Sara seems to have a moment of clarity and starts working on a somewhat fluff piece for the magazine. Neil walks away from Sara's desk, quietly, with his coffee, scanning the room to see if anyone had been listening… Neil was a player.

Sara, over the several days, is working on her next piece, not very satisfied with it at all. Worry about her job is settling in again, when her boss calls her into his office. "Morning, Sara. Listen, I know you're having a hard time finding something solid to work on, so I just came up with a piece for you. We do like your writing style and this is something you can sink your teeth into. Also, this is something the folks upstairs really want done, so……looks perfect for you." "Oh…okay," says Sara. "What is it exactly?"

"It's a piece about government corruption across the board these days, with a primary focus on the postal service. You know, being used to affect voter access, deteriorate its own performance, in an effort to get rid of itself and quash the voice of people and decrease government spending to increase the size of the bulging pockets for those at the top. You know, fascism and that sort of thing. The people upstairs at this magazine think the public is clamoring for something like this right now. Certainly, shouldn't be difficult to dig up some real dirt on this. You'll probably need a bulldozer. Ha ha. You can requisition one of those from the finance deptartment. Ha ha. So…what do you think?" asks the chief editor.

"I think it sounds…fine. Yes, I'll dive right into it. Thanks," says Sara, not really all that enthusiastic about the idea. She goes back to her desk, puts her fluff piece in the bottom drawer of her desk and begins some online research, with a big sigh. Two days later, she has a renewed interest in the climate game story.

CORPUS-A V2.0 has a fund of its own. It's based upon a blockchain concept, where players would pay the minimal annual subscription fee, which would be invested largely in cryptocurrencies. A player pays their annual subscription fee, which is invested in crypto after a minimal percentage is deducted and placed in a separate crypto account owned by the game company, Harthun-Broadmoor. This covers company expenses and profit. The rest, deposited in the Corpus-A game fund, is available to pay rewards, minus a small fee, to gamers once they prove they have solved a climate problem. They may collaborate with other players and share rewards.

None of these actual inner workings of CORPUS-A are known to Sara, yet. She notes an article online about the disappearance and assumed murder of a climate activist in western Europe. Reading through the story details, she finds that the activist was involved in playing the game and she wonders if that is connected to their death. Putting all the fairly recent positive and negative events together, Sara figures this just can't be a coincidence. What's behind this? Is it that activism is becoming more ingrained (which she figures is a story in itself) or is this game somehow related? She digs deeper.

It's the winter of 2020. Climate enthusiast and activist, Arianna Romano, has emerged, as a young Italian climate figure. She graduated from college with a degree in environmental science and still, at 23, lives with her family in Milan. She writes articles for local papers on ways to live more sustainably and the need to improve the climate by curtailing the use of fossil fuels and greatly expanding the use of green energy. As a so-called activist, she is little known, even in Milan. Out with friends, one evening, she is exposed to CORPUS-A through casual conversation with friends, one of whom is an active player and "just loves it." She agrees to look into it and the conversation switches to a ski trip they have all planned. Everyone is excited and looking forward to it. They are all pooling their money to hit a really fashionable resort near Buccinasco, just outside Milan. Madonna di Campignoli is a high-end ski resort, where the upper echelon goes to see and be seen. Skiing is a secondary consideration for many who go there. Ari and her friends hope they will see some celebrities…and get in some "turns."

Arriving at the resort, they put away their gear and immediately hit the streets for shopping and people watching and then the bars for the night life. The next morning, they all drag each other out of bed to get ready to ski, still being hungover from the night before. They quickly revive, however, after getting on the chair lift in the freezing cold air. The skiing conditions are really very good and Ari is an accomplished skier, better than most of her friends. The group decides to take an intermediate run, which is wide open and somewhat crowded on the weekend. Ari suggests a better run, which is a little steeper and more isolated, but more picturesque. All her cajoling is to no avail and she agrees to meet them at the bottom of the gondola for their next run. "You're all going to miss a great run," says Ari and they all depart.

Ari is never seen again. Her friends are unable to contact her, on her cell, and they wait and wait at the bottom, prior to contacting the ski patrol, which then sends two patrollers up to look for her. The patrollers take the same run down as Ari and find nothing. A couple more patrollers are sent up to aid in the search. After an hour or so, a town rescue team is alerted and they also join the search, this time with two rescue dogs, who quickly divert their attention into the woods on the side of the run. A patroller and a rescue crewman go directly to the spot where the two dogs are digging, as if there was an avalanche… There was not. They find a small amount of blood, which they initially think was from one of the dogs' feet, from digging; however, after inspecting the dogs, they realize it was not. The two rescue people look at each other in a quandary. What was this? Did the young skier hurt herself in the woods and ski away down to the resort clinic? No tracks indicate this and a quick radio call to the clinic rules that out. There are many ski tracks and footprints in the area where the blood was found. What went on here? An uncomfortable silence befalls the two rescue personnel before one whispers, "Uh-oh, let's get down and notify the Policia di Buccinasco." The two depart and the rest of the crew continue to search other parts of the ski run. Ari is not found… ever again. As expected, her friends are distraught and look for her for the next two days. She is not found. They return home to visit her parents, who are out of their minds with grief and her father is so angry he starts saying that he has "connections" and he will find out who did this and they will pay. It turns out that Ari's father does, in fact, have remote family connections to the Mafia in Sicily. His wife, Ari's mother, begs him not to get "that side" of the family involved, but Ari's father is so upset all reason has gone out the window and he places a call to a distant cousin. After some meaningless chatter, he describes the reason for his call. His cousin tries to console him and strongly suggests that there is no family Mafia connection and that it's all in the movies and he's reaching out, just because he is so understandably upset. Ari's father gets angry and states he knows that there is family who can help him. His cousin states clearly that there is no such thing and even speaking about this, over the phone, can get people in trouble and bring a bad reputation to those who don't deserve it. His cousin suggests they get together for dinner because they "haven't seen each other in so many years and they should reconnect." They set up a time and place to meet.

CHAPTER 7

Pandora's Box

Sara is back at work and her investigation of the Ariana Romano disap-pearance in Italy comes up empty. A complete dead end. No specific police records, nor any suggestion of Ari's whereabouts. No further reports about her, at all. Now, Sara is, after all, an investigative journalist and this and prior events that she has looked into have a peculiar pattern she has not seen previously. All have a complete hard stop with no follow-up information. The events all seem to be environmentally and climate related. There does seem to be some correlation with the CORPUS-A game, or so she surmises. She decides it's time to pay a visit to Harthun-Broadmoor Entertainment, the publishers of the game. Sara makes an excuse to her boss that she is ill and is having some tests done that she is worried about and will need a few days off. Her boss is a bit frustrated but does have compassion and advises her to take the time she needs and to stay in touch with him. She agrees and makes plans to go to Denver and visit the game company.

Sara, in her planning, finds that the designers of CORPUS-A are Hudson Harthun and Jamie Broadmoor, the CEO and COO of the company. They agree to meet with Sara, who states she is doing an article about climate change and is very interested in their game, which her nephew plays and from which he has learned "so much." She also states that she is so delighted with his intellectual growth and level of deductive reasoning at such a young age.

"He's like a miniature Sherlock Holmes now, much to his parents' level of annoyance at the dinner table. At his age, I was playing with dolls and trying on my mother's high heels. He makes me feel like kind of a dope now." She laughs. "I am sure it's due to your game."

Sara meets with Hudson and Jamie in their beautiful office, and she is escorted by Jamie's executive assistant to a conference room, one side of which is all glass and facing the mountains to the west. She is offered coffee or water, but declines. Hudson arrives and introduces himself. He is well dressed, but trendy casual and very cheerful, as he glances out the window. "Beautiful, right? I could spend all day, in here…and often do," he says, laughing. He sits down across from Sara and again asks if she wants anything. Sara politely declines. They start some small talk when Jamie walks in. Hudson introduces her as the COO, "the one that makes the game so educational and intriguing."

Sara is stunned by Jamie's appearance, as most people are. A beautiful, young woman, impeccably dressed and mannered, not looking like someone who is used to rolling up her sleeves to slug it out with some fancy computer coding. "Hi, Sara. Very nice to meet you. Pay no mind to my partner, Mr. Harthun, who is the real brains behind the game concept for CORPUS-A."

Sara explains how she had a mental image of a couple of scruffy young people with headphones around their necks, looking like they just left mom's basement. "I apologize for that stereotype. My gosh, I'm experienced and old enough to know better," says Sara.

All three laugh and Hudson relates, saying, "I looked like that just a few years ago until my partner told me to buy a suit."

Jamie quickly adds, "He's not joking, you know…Men!" They all laugh.

Sara is, once again, struck by the pair. "They seem so close," she thinks. "What's that about? Are they secretly a couple or related somehow?"

After the opening pleasantries, Sara gets down to her interview, which Jamie and Hudson feel will only be good and free publicity for their game. Jamie and Hudson lay out how the game comes to be so educational, with occasional subliminal messaging, to enhance learning and motivation to learn. Jamie also relates how each version of the game incorporates climate events that have occurred since the last version.

"Oh," says Sara, "so the game doesn't incorporate new events?"

Hudson looks at Jamie and says, "Uh, no…how could it do that? No game system can really do that, Sara. The sheer processing power required to do that would be enormous. As it is, the game has to put forth problems for each player based upon their profile and game successes and failures. It's tailored to each player, you see. It has to assign rewards in monetary terms and status icons and rerun their updated profiles, through really complicated algorithms, before it dumps the data into a blockchain-like system for a payout. Believe me, that's an enormous task and even that is only feasible because of the fancy computer chips designed by my partner."

Sara is trying to take notes and, at the same time, wrap her head around what she is being told. "So," interjects Sara, "there is no way the game could be related to very recent global climate events, such as the disappearance of people who are climate deniers or fossil fuel producers, for example, who are just greedy and not climate sensitive?" There is a silence.

"Sorry, I'm not sure exactly what it is that you're asking, Sara," says Jamie. "This is a game…right? It's virtual reality, not reality itself, right? Of course, we've noticed these world events that you speak of in the news, but…not sure what that has to do with our game," continues Jamie.

Sara continues, "Exactly how much do people get paid for a successful solution to a climate problem and how does the system know it was a success?"

Hudson speaks up, saying, "The amount of cash reward…not really cash, per se, but cryptocurrency, is based upon another complex algorithm judging creativity, probability of success, severity of the initial problem, etc. It judges success very simply. When we design each game version, we already know, as does anyone, who watches the six o'clock news, what happens to climate related problems. If they're resolved, the game knows that, internally. If the problem is never resolved, the game pays off based upon probability of success and that is really getting deep into the algorithms, which, no offense, is really complicated math. Nevertheless, Sara, each game version is based upon yesterday's news."

Sara continues to dig, "But don't you both find it weird that there are all these, for lack a better word, 'dark' events happening over the past couple of years to these fossil fuel people?" "Well…yes, frankly," responds Jamie. "But we attribute it to a growing awareness among young people of their bleak future being created by these people. Don't you want your niece and nephew to grow up in a world with clean air, clean water, plant life and daily temperatures less than 115 degrees…without wild fires, floods, tornadoes and droughts all year round? So, again, Sara, are you trying to create a connection between sinister world events and our 'virtual reality' game, which is purely educational and based upon old news?"

Sara, uncertain about how much to push and how much tech she would understand anyway, says, "Oh, no…I'm sorry if I gave you that impression. I just find these recent reports so disturbing. I'm just trying to understand it all. It's just…what I do."

Hudson now abruptly brings this brewing argument between the two women to a hard stop. "It's been nice to meet you, Sara. We hope you have enough information for your article. Good luck with it. Jamie and I now are about five minutes late for a software maintenance update meeting, so we really do need to go. Susan, Jamie's assistant, will show you out. Bye."

Jamie and Hudson leave the room and take the elevator up to Jamie's office and close the door.

"So, what the hell was that all about, Hud?" says Jamie, obviously upset. "That woman obviously had an ulterior motive; an agenda that we were not initially aware of. She wasn't just looking at CORPUS as a game. She was looking to dig up some dirt and if she couldn't find it, I'm sure she'd make it up!"

"Whoa," says Hudson. "She's a journalist. They dig and dig to get a 'big' story."

"Yeah, and they make it up, if they have to," says Jamie.

"Well…it does seem like she was jumping to an outlandish conclusion, all right. She clearly doesn't have much tech education or knowledge, so it does seem like she interprets things in a kind of ignorant way," says Hudson.

Jamie agrees and they settle down a bit. There's a brief silence, with Hudson fiddling with a paper clip on Jamie's desk. "Just between you and me, Jamie, do you think there's even a remote possibility of a connection here? You know…between some jackass deep pockets fossil fuel king and his disappearance…or whatever?"

"Okay. I admit it has crossed my mind, but even really stretching tech reality to a ridiculous degree, I can't imagine how any of this could be related to our game. We just didn't design anything like that into it. How could we?"

"So…what…the game is running wild on its own?" says Jamie.

"Yeah, it's da Terminata," says Hudson. "Everyone knows that's completely real." They both laugh. Hudson then asks Jamie if, during her biochip research and design days, she ever saw any evidence of automated or self-learning or self-awareness of the chips.

Jamie responds with a bewildered, "No. I can't even imagine how that would happen. These chips are sophisticated, as you know, but they're not brains. They just have the speed of silicon with the compactness of biological nervous tissue. So, very fast and much more processing capability, but they are still running a canned program, right? They use old information and determine the initial probability of a solution, already knowing the outcome."

"Yeah, yeah…I know," says Hudson, "but maybe you and I should start playing the game for a couple of weeks to see if either of us can find any bugs. You know, do our 'due diligence.'" Jamie pauses, staring at Hudson. "Okay…yeah. What can it hurt? What the hell?"

Meanwhile, in a quiet cafe in Rome, Ari's father is meeting with his cousin from Sicily, having not seen each other in years. Her father is still somewhat distressed, but after a glass of wine, he settles a bit. His cousin is quite relaxed. The two exchange pleasantries very briefly before the pressure of the business at hand overwhelms Ari's father and he, again, tries to pull information from his cousin, whom he still believes is connected to the "family," i.e., the Mafia.

His cousin leans in and whispers that they need to keep the volume down saying, "You never know who's listening." His cousin then opens up, a bit. "Listen, I told you I would check with some people for you about your daughter, and I did. I found that she is alive and living in an uncertain location, for a while at least; it's uncertain to me. I don't know where she is, but I know she's in Italy and is okay. Here's what else I found—" Then he stops and tells Ari's father, "You do know you can't repeat this to anyone, and I mean anyone, no one, not even her mother, you understand me?" Ari's father agrees. "Okay, so the people I spoke to told me there's this video game, you know, like the little kids and the useless older kids play all day. Well, word got around that one of these older kids got too big for his britches and based on what he saw in that game, found out that this young woman, your Arianna, I mean, found out there was a lot of money, and I mean A LOT of money to be made in this game, doing…let's say, unsavory things.

"Your Arianna, she's a good kid and didn't like what she was seeing and decided she'd blow the whistle on the game. Apparently, she was on one of these blog things and made a comment about this to a friend. Next day, boom, she was grabbed. Now, the grabber's father found out about all this and punished his son, pretty good, but he also looked into this game and brought the info to some 'upper level' people, who decided they needed to be in this game, too. So, this brings us to your Arianna. The people who have her are…you know, family people themselves. They see she's a good kid and just want to be really sure she says nothing of any of this to anyone… anyone…and I don't need to tell you again, either, right? You'll get your daughter back."

"Wait," says Ari's dad, "there was blood where she was taken?"

"Yeah, yeah…this guy's kid, you know the one who grabbed her, popped her in the mouth when she screamed. No big deal. She's fine and the guy's kid got his, believe me. The damn fool kid exposed his father to an unwelcome level of scrutiny, in his town, for awhile. Now, go home to your wife. Just tell her everything will be fine. Reassure her but say nothing else, nothing, no details, at all…again, you understand? Just smile and say everything will be okay very soon. Hey, say hello to the misses, okay? We should all get together for the holidays or something…not wait, what, another 10 to 12 years, is it?"

Ari's dad leaves, thanking his cousin profusely and feeling much much better. He can't wait to see his daughter, although he knows he is going to get an endless third degree from her mother.

CHAPTER 8

Revelation

Jamie and Hudson are each at home playing their CORPUS-A game after office hours. They do this for a couple of weeks whenever they get the chance, and the next day they touch base in the hallway and say, "Nothing to report," and they chuckle. One particular morning, however, Hudson sees Jamie in the hallway and she quickly and predictably says, "Nothing to report." Hudson suggests they go to his office for just a minute to discuss something. Jamie looks at him with a curious expression on her face and they proceed to the elevator to go upstairs. Once inside Hudson's office, with the door closed, he reports that he came across a novel game problem and he has a list of possible solutions. He wants to see if Jamie knows anything about it or if she has seen anything similar. Jamie asks him what exactly he was seeing.

"Well," says Hudson, "I was playing last night and everything was normal and predictable. Then, after solving a quick problem in the usual way, I was offered a new problem, one I had just read about in yesterday's morning news clips. The operative word there being 'yesterday.' I was a little startled, to say the least. Then, it offered me some solutions that were 'creative,' to say the least, and then the usual offer to custom create my own solution. I put in: 'permanently get rid of that guy.' Probably should not have done that, but I just had to see what the game said next. It asked for details, so I put in seriously sinister stuff and CORPUS came back saying: 'Excellent choice. High probability of success. Will let you know your outcome soon. Please keep checking your account.' Jamie, I nearly fell out of my chair. Felt like I was being punked and I should look for cameras spying on me. Really, felt a chill go down my spine! Do you have any idea…I mean ANY idea, how this could even be remotely possible?"

At this point, Jamie had pushed herself way back in her chair, almost like she had just seen a shocking event in a horror movie. "Are you joking? No…of course, you're not. I can see it on your face, Hud. What the fuck!!?"

"Yeah…I'll say. What the fuck!!?" says Hudson.

Jamie follows up, saying "I…..I don't……uh…..this can't…. whaaattt the fuuuucckk?" She adds, "Is someone fucking with the game in the software maintenance or security departments? We need to immediately contact the chief architects in those departments and get to the bottom of this, right now!"

"Hold on," says Hudson. "Do you know what you're saying? Think. Think for a minute. If this is a platform wide issue……my god, a global issue, can you even begin to see our liability? Oh man, oh man, oh man. We have to be very careful who we speak to and what we say and we need to carry recorders for every conversation we have and tell them we are concerned about our 'bottom line,' not something so much more sinister. Let's go get some coffee, not in the office. Let's go to Ruurlo's. Haven't been there in a while." The two head to the elevator.

At Ruurlo's, they sit in their usual booth and the usual waitress comes right over. "Hey, you guys! Haven't seen you, in a while. Wow.

Lookin' snappy today. Ooh, you guys look like money today. Gerrit, come over here and say hi."

The owner of Ruurlo's comes right over. "Heyyyy. How are you guys? Been a while. Oh, sorry, let me go change into a suit and tie. Didn't know we were gonna see computer game royalty today. Ha ha."

Jamie says, "Hi Gary. Ha ha. We're the same old nerds...well, I am. Hud is feeling highfalutin these days, maybe." She chuckles.

"Hi Gerrit," says Hudson. "Highfalutin? Me? Look how she's dressed." Hud points to Jamie. "If her heels get any higher, she'll be falling forward, in the hallways, all day long, ha ha."

Gerrit and the waitress laugh and then Gerrit leaves. "Great to see you guys again."

"You too, Gary," says Jamie.

"Jeez, it's Gerrit...Gerrit," says Hudson.

"Don't worry." Gerrit laughs, walking away. "I even answer to my wife when she says, 'Hey, you. Get over here!'"

Coffee, scrambled eggs and yogurt arrive and the two settle down. No one is sitting near them. Hudson starts, "So, let's get down to it in a systematic way. We have a game. It's programmed to use past climate events that WE have put into the database, where the current status or outcome is already widely known. The players put in their best guess solution, and the game calculates the likelihood of success based upon the actual outcome which, again, is already known and an algorithm that calculates probability of alternate outcomes. Pretty straightforward. So, how could it know about an event that we didn't program in...and... and, how could it come up with possible solutions to such a problem or the probability of success? I have only one possible answer. How about you?"

Jamie sips her coffee. "No, I agree. I'm thinking there is also only one possible solution. Someone, unbeknownst to us, has to have added something to the program. It has to have occurred at the maintenance department level since millions of copies of CORPUS are already out in circulation. Also, someone must be doing it on a regular basis and keeping it updated. And… and, that someone must be adding possible solutions, of their own and changing the algorithm to calculate probability for players' own custom solutions. How could they do that, though? I mean, we were being really, really innovative when we came up with these algorithms, weren't we? Looks like we have someone on staff who is just as innovative."

"Yeah," responds Hudson, "and a real psychopath, to boot." The two look at a list of company employees in the maintenance and the architecture departments. No one's name jumps out at them at all. "Can we ask HR to do a background check on few people, you know, quietly?" asks Hudson. "We could maybe ask your mom," he adds.

"I don't know," says Jamie. "Is that legal and ethical?"

Hudson states he thinks they're way past "ethics" at this point. "We really HAVE to know, and soon."

"I agree. I'll ask her." Jamie places a quick call to her mom. "Hey, mom. We have a company concern and a question for you. If we give you a few names in the upper levels of our staff, can you run a complete background check on them…and I mean a complete background check…everything there is to know about someone?"

Jamie's mother states that the person has to know that background checks are being done, but adds that when they were hired, they knew a background check would be done, so the company could just say this was part of that if the company ever needed to justify it. Then she asks if this is really, really important because this is a gray area.

"Yes, mom. This is really important. High level stuff. Please don't even discuss this with anyone. I'll let you know, later, what's going on."

Jamie reads a short list of the top three people she and Hud think could even possibly be smart enough to pull something like this off. Then they feel terrible about it since they know these people well, or thought they did, and like them very much.

"You can never really know someone, though, you know," says Hudson. Jamie agrees, but looks sad, as she sips her coffee. Hudson recognizes this. "Yeah, I know. I feel terrible too, but we're trying to be good guys, right? If someone is messing with our game and some-one is getting hurt…to hell with the money or the company, for that matter. This is way bigger than that."

He sees Jamie's eyes getting glassy, tearful and he does his best to console her. The two know each other inside and out. This has nothing to do with intellect. This is emotion that can only come from unspoken communication and a quasi-family relationship. It's quite clear, the two, while not biologically related or married, are inseparable. When one is sad, the other is sad also. It's empathic.

CHAPTER 9

The Usual Finger Pointing

It's clear that the employees Jamie and Hudson are considering possible suspects based upon their expertise in and experience with very sophisticated game concepts are people they would never otherwise suspect. These are people whom they have invited to their homes and have celebrated with during office parties; they all have families. How could they be involved in this sordid situation? Hudson and Jamie are secretly hoping that a thorough background check comes up empty. "But," wonders Hudson, "what do they do then? No one else is capable of modifying CORPUS like this, so…."

Jamie and Hudson are trying to keep it "business as usual" at work by compartmentalizing the whole issue. (Hoping it will magically go away. It doesn't.)

Five days later, Jamie's mother calls. "Hi sweety. It's mom. I have some information for you…about what we discussed several days ago."

"Oh, mom! Let me get to my office." Jamie stops by her assistant's desk and asks not to be disturbed by anyone other than Mr.

Harthun if he should stop by. She goes into her office and closes the door. "So…mom…I'm hoping you found nothing suspicious about these three people although Hud and I don't know what to do next if you didn't. It's freaking us out."

Jamie's mother goes on to explain about background checks and how even the best of them do have some limitations. There can be records that have been sealed, for various reasons, issues that have never been reported, by schools or companies, etc., because of possible retaliatory litigation or just good will when companies part ways with staff, etc. "So," says Jamie's mother, "two of the three employees seem to be spotless, even better than I would ever expect…I mean people get dinged for tax issues, nonpayment on credit cards or mortgages or car repos, but these two people have nothing on their reports. Now, the other one, Mr. Abeer Gupta does have some things of note. After less than a year, he was apparently released from or quit his last position in IT. There is no report about why, which is really not uncommon, but less than a year? What's that about? Also, he apparently had difficulty, just a few years ago, paying off a personal loan for $100K, and it doesn't look like it was related to a mortgage. He was also employed and making about $175K at the time. So, you can take what you will from that issue, but it does look like he needed 100K for 'something.' He has been with you from the start of your company, and I actually think I remember meeting this man at a company holiday party at your home, not that that means anything, of course."

"Well," says Jamie, "I don't know where to go with this. It could be nothing or it could be something. What should I do?"

Her mother tells her that she could have a private investigation done on him, but when it comes to IT, this is very difficult because the investigators are usually not very good at it. She also tells Jamie that the best general advice is to keep her friends close but her enemies closer. She advises that they start a company review of several departments, including Mr. Gupta's, under the guise of making sure the company is meeting their subscribers' needs. Then, they can ignore the reviews of all departments, except Mr. Gupta's and can keep a close eye on him for a while. She suggests six months to a year. If they don't find any issues, she suggests stopping the review. If they find company issues during that time and are certain he is not involved, they need to be creative, be daring and look elsewhere. Jamie thanks her mom, who asks if Jamie wants to tell her what this is about, but Jamie tells her that she will, but later, maybe.

Jamie calls Hudson and tells him to come to her office. Hudson arrives and they sit down, with Jamie describing her mother's findings and advice. "Hmm," says Hudson, "it doesn't sound like much, but yeah, let's go with it. I'll set up the review, but six months to a year? I don't think we can just sit back and watch this. That journalist we met with isn't going to let this go and neither should we. We should do other things at the same time as this review."

"Like what?" says Jamie.

"Let's go back and look at the programming and how it's evolved over the last couple of versions. I can do that. Let's also look at how the chips you designed are functioning. Maybe we can take some chips out of the game server we've used for a couple of years and test them for errors in processing or something. Can you do that?"

"Looking for what…what?" says Jamie.

"I don't know," says Hudson. "That's your area, but I think we need to look at everything we can simultaneously, don't you? I mean, we both want to find out our company and our game are squeaky clean, right?"

"Yes, absolutely," says Jamie. "I'll drag out my old papers on the BioSils." (This is the patented name for Jamie's processors, which are part stem cell generated human neurons on a silicon base. This makes the processors fast but also very small, packing in much more processing power in a smaller chip footprint.)

The two start the next morning on their tasks. Hudson goes to the software maintenance department, announcing he will be doing a systematic review of his game program from the graphics, to the user interface, to the game mechanics. He will also be looking into the blockchain type player profiles, solution and reward system and the players database (which does not contain any players' secret keys nor any personally identifying information). He gets some quiet and strange looks, from the programming architects to the entry level software developers. Hudson then, laughingly, says, "I see some strange looks. Not to worry, I'm only doing this because our legal department has suggested I not distance myself too much from the design and maintenance people, to keep the game consistent with our players' needs. I'm not a spy and don't care if you have food and drink at your work stations. Wait, when I say drink, I mean soft drinks. I actually DO care if you're drinking something harder......and you're not sharing with me." He chuckles. "Really, just go about your daily routines and pretend I'm not here...actually, I mean that. I want to get this done, so please don't interrupt me unless it's an emergency."

Jamie spends several days, digging through her old biochip research papers. Seeing nothing weird or even remotely interesting, she takes her most important research test results with her and proceeds to the chip processing lab. This is not a typical cleanroom, where all the people inside are walking around in pure white bunny suits with shoe covers and paper masks, the ones where to enter you have to go through double doors with laminar air flow, all designed to keep the areas dust and bacteria free. (These can short circuit the fine printed circuits on a photographed chip.) No, this is a level 3 biological cleanroom, where silicon chips are made and overlaid with stem cell generated human neurons. Electrical impulses come into the chips, are "processed" and fed into a matrix of human neurons, which are much smaller. Most of the impulse direction or "processing" is done here. The output from the neuron mass is cleaned up by the silicon base again before being distributed. The output is predictable and expected, as it is in a pure silicon chip.

Jamie dons the usual bio cleanroom uniform and goes inside and takes an empty workspace, off to the side. A supervisor immediately comes over and asks her name and why she's there. When she looks up at him, he looks at her face and can then also see her badge.

"Ms. Broadmoor?" he exclaims. "I … I wasn't expecting you at all. Sorry I didn't recognize you, but you know—"

"Yes, it's fine. Robert, is it?"

"Yes, can I help you with anything … anything at all?"

"Yes, Robert, if anyone asks, I'm just here doing a thorough review of my chips. Mr. Harthun and I were advised by our legal team to do periodic reviews of our software, which he is doing, and our hardware, which I am doing. I'm going to be here with you for a while, and I would just really like not to be disturbed because I'd really like just to get this done. There are many other things I'd rather be attending to right now, believe me."

"Yes, of course. I'll be sure you're not disturbed and if anyone does bother you, please let me know."

"Great. Thank you, Robert." Jamie begins work.

Jamie, a very methodical, systematic thinker, starts by running a few tests on some brand-new chips. She finds nothing unusual. Then she requests some old chips that were removed from some of their older servers. These are kept for study down the road to improve the design, but this is costly and has not really been done yet. She specifically requests chips that have been in use for at least 18 months. These are brought to her and she begins bench testing them. They appear fine. She tests quite a few over the next couple of days. Nothing unusual.

At this point, Jamie is getting frustrated and feels she is wasting her time, doing … what? Looking for … what? She decides to go back and test the older, "used" chips, with a "game load." They all seem fine in a standard bench test, where there is expected output, but what do they do with a new problem that she dreams up out of nowhere? The output should just be blank or incorrect or just a garbled mess since the chip has not seen that problem before. Instead, she gets a series of possible, even plausible, solutions and the standard opening for a "custom" solution. She is stunned momentarily, then collects herself and does a retest. She gets the same output. She puts several other "old" chips through the same scenario and gets garbled, meaningless output from most of them, as she would expect … but not all of them. Some also create custom output. Jamie takes the old chips that are creating the unexpected output, puts them in a special container and quietly leaves the lab.

Once out of her lab attire, Jamie takes herself to Hudson's office, calling him on her cell along the way. "I think I have found a possible answer to our predicament," says Jamie. Hudson starts to speak, but she cuts him right off, telling him she is on her way to his office right now so that they can discuss it in detail. Both Jamie and Hudson are new to this cloak and dagger style of communication and behavior, but both realize the implications of their investigations and these weigh heavily upon them. This is not who they are, nor is it who they wish to become, but they understand that a little caution is very wise right now.

Arriving at Hudson's office and closing the door, Jamie clears a space on his desk quickly, opens her brief case and puts down a container containing the suspect chips. "Know what these are?" asks Jamie.

Hudson looks briefly at them and says, "Yes, are you serious? They're our specialized BioSil chips. So, what of it?"

"Well," says Jamie, "they may look like our chips, but they sure don't act like them. The BioSil chips don't look like normal computer chips. They're not just square or rectangular pieces of plastic with metal contacts along the edges. These chips have a biological component and, as such, they need to be kept in a nutrient bath or gel that is constantly replaced. This keeps the neurons alive and functioning normally. They do have metal contacts around them, but they also have small ports where nutrient fluids can be 'circulated' and kept fresh. Their temperature needs to be maintained properly also or they may malfunction."

Jamie goes on in detail with Hudson, about how the chips are built. She describes the standardized amount of biomass in each chip. Then she tells Hudson how in a standard bench test, which was done in a copy of the CORPUS game not connected to the internet, these chips performed fine when she forced them in one of their standard game problems. She explains that the chips then gave the expected output. When she put in a custom problem, however, instead of coming up with some gibberish, the chips came up with an array of never- before-seen, possible solutions with space for the player to come up with novel solutions. She tells him how she would pick one of its solutions or even put in a custom solution of her own and the chip would calculate probability of success!

"What?!" says Hudson. "How? Did you test other chips?"

"Yes," says Jamie. "I've tested all these you have on your desk right now, but listen, I've got to get these back to the lab and reconnect them to the fluid nutrient lines or they'll die, and I won't be able to test them any further."

"Well, what are you going to be able to test them for now?" asks Hudson.

"I'm going to consider putting them into a live game console, connected to the internet, and playing a game based upon either your or my profile as, hopefully, the highest level players," says Jamie.

Hudson sits down, at this desk, thinks for a bit and then asks Jamie if she thinks that's wise. Jamie responds that she feels it's their only option and they don't really see the whole picture yet. She also tells Hudson she thinks it's safe because of the blockchain-like security. No one should be able to identify them.

Hudson has the facial expression of a five-year-old who was just asked to spell, syringomyelia. "Okay, then. Should I go with you? I'm not sure what I could help you with, but… hey, use my profile in your testing. I'm a bit better player than you, right? Let's see how it does when I'm the player."

Jamie agrees and tells Hudson he does not need to accompany her yet and it would look really suspicious if they both walked into the lab together. Everyone would know something was up, and they might think they were all getting fired or something.

Back at the lab Jamie reconnects the old chips to the storage nutrient source. She keeps one chip out and before she installs it into a live internet connected game console, she tests it for size, weight, contact cleanliness and temperature. She notices a peculiar tiny increase in the weight of the chip. It's not quite within tolerances. She thinks this may be due to some sludging of the nutrient fluids in the chip, and she installs it into the game console. Then, she identifies herself as Hudson, using his profile. After playing for about an hour with the highest level climate problems, she comes across a problem which she recognizes as "new." She recalls Hudson talking about it after seeing a news clip, just a few days ago. If she hadn't been wearing a bio cleanroom mask, her shocked expression would have been noticed by everyone in the room.

"How is this possible?" she thinks. She requests a menu of possible solutions and they pop up, clear as day, with probability of success numbers. Looking around the room to be sure no one is watching her play, she then enters a custom solution after giving it a lot of thought… a lot. She enters: "Find the person responsible and shoot them." The game instantly calculates the probability of success, which is higher than all the other possible solutions. The game then requests: "Method Details?" Jamie shuts the game session down, promptly. She removes the chip and stares at it for a bit.

Seeing her staring at a chip, the supervisor comes over and says, "Everything okay, Ms. Broadmoor? Did you find a chip problem?"

Jamie is briefly startled by the supervisor, but replies calmly, "Oh… no. I'm just daydreaming about something completely different."

"Yeah, ha ha. We all do that," replies the supervisor. "Let me know if you need anything."

Jamie thanks him and when he leaves, she segregates and marks the chip, so she can find it later. She goes home for the day, planning on coming back after hours, this time with Hudson.

Once back in the main admin building, Jamie goes back up to Hudson's office and walks in. Hudson is having a meeting with a prospective upper level manager. Jamie excuses herself for barging in and apologizes, but tells them both that she has some new sales info that can't wait and asks politely if they can reschedule their meeting. The interview closes with the potential new hire saying that's not a problem for him and Hudson saying that the interview is going well and he'll be in touch later in the day.

Once alone, Jamie tells Hudson that they need to go to the lab tonight to do further investigation into this chip and she explains why. The stunning never seems to stop, with Hudson speechless after listening to Jamie tell him what she found. He no longer asks "why?" or "how?"; he just agrees to go to the lab. Later that evening, they find the answer and while everything is now clear, a finished picture has emerged. Jamie has opened the BioSil chip and has found that the neuron biomass has gotten slightly larger. This accounts for the weight difference. Jamie surmises that the neurologic material has started dividing and, even more unsettling, has mutated in a manner that now allows it to randomly sample real time data bases, such as news media, determine if they meet the standard game criteria for a significant climate problem and postulate possible solutions tailored to the sophistication of the player profile. She tells Hudson that her best guess is that the constant electrical stimulation from the silicon part of the chip has changed the neurons' DNA allowing it to replicate and giving it a level of artificial intelligence, not previously known.

"Is it really thinking?" asks Hudson. "Is it alive, so to speak?"

"No," responds Jamie. "I don't think so. It just can handle so much more data than I ever expected. It can obviously extract all sorts of data from the internet and come up with really novel solutions."

"Yeah, sinister stuff!!" says Hudson. "What's the end game with this thing?" he asks.

"I don't really know," says Jamie, "but it obviously can't grow much more because of the size of the chip container. I think it will try but then will crowd itself, and this will mess up the synapses and the output will start to be corrupted. Ultimately it will just be gibberish."

"You think… but you don't know," says Hudson, "and it really doesn't matter. Look at the problems it has already created. Unscrupulous players have been 'taking people out' and getting paid really big bucks for it! We're liable for this! Can we say we didn't know and couldn't have known because this is cutting edge technology? Should we shut down the game completely? Should we go to the press? Oh… those poor people! We did this, Jamie!"

Hudson is starting to hyperventilate. Jamie grabs his shoulders and helps him sit down before he falls. "Listen, Hud, I know you. I love ya," Jamie says with tears in her eyes. "Look, you're a good guy. You mean well. So do I. We couldn't have known. I myself didn't ever envision anything like this. I feel just like you do right now. I don't know what to do, but that fact makes me and should make you ultra-cautious. We can't just fly off the handle. People will not understand. We need a well thought out, reasoned approach. Breakfast tomorrow at Ruurlo's and business as usual in the meantime, right? Don't you think?"

Hudson agrees. "Uh, do you have anything you ever use for sleep? I take some of my mom's stuff, occasionally, if I can't settle my mind and get a good night's sleep. Want me to stop at mom's and bring you one? I'm definitely going to take one," says Jamie. Hudson agrees and actually leaves his car at the office and goes with Jamie to her parents' home, where they both take a "sleeper" and stay in separate rooms.

CHAPTER 10

Wait... Who Are You Again?

The next morning, Jamie and Hudson are oversleeping. Jamie's mom finds Hudson on the couch, fast asleep. She goes to Jamie's old room, now the guest room, and finds her asleep as well. She starts speaking to Jamie in a soft voice, "Jamie, are you awake? It's after 9. Why is Hudson in the living room on the couch? Did you guys party last night or something?"

"What…. no…no. We have a meeting this morning and needed to be fresh, so … sorry, we each took one of your sleeping pills. Ugh… I feel like I was hit by a train."

"Dear," says Jamie's mom, "you shouldn't do that, you know. Who knows? You could have been allergic or something, or Hudson could have. I should go wake him up. You'll be so late for your meeting."

"No, mom. It starts when we get there," referring to the two of them just going to Ruurlo's. "I'll get Hudson up, mom."

Jamie gets out of bed, throws on some sweats and goes in to wake Hudson. "Hud… Hud, get up."

Hudson grumbles and then starts to open his eyes. He looks up at Jamie. "What? What are you doing here, Jamie?" He looks around the room. "Wait. Where are we?"

"We're at my parents' house. Remember, we took a sleeping pill and now it's a quarter after 9. We need to get to Ruurlo's and organize our thoughts. I'd say we both slept pretty well."

"Yeah," says Hudson, "pretty too well."

They both get their clothes together. Jamie's mom offers some breakfast, or at least coffee, but they decline and after collecting themselves, get into Jamie's car and drive to Ruurlo's. By the time they get there, they're more awake and feel more settled. Good thing because this will not be a stress-free day, by any means.

Upon entering Ruurlo's, they are not met by Gerritt or their usual waitress. They find that a bit odd, but proceed to their usual table. Their waitress finally shows up and just says, "Coffee?" No smile, no small talk, no joviality, just "coffee?"

They look at her and both say, "Yes, please."

Hudson looks at Jamie and whispers, "I think we have just entered The Twilight Zone." Jamie replies, "Yeah! What's that about?"

Shortly, Gerritt appears, walking over to their table with a slender, well-dressed gentleman. "Morning, guys. Been waiting for you," says Gerritt.

"Really, Gerritt? How did you know we'd be here this morning?" replies Hudson.

"I have a gentleman here who really wants to meet you, so I'll leave you to it." Gerritt walks away before Hudson can even say anything else. The gentleman smiles and sits next to Jamie, sort of pushing himself into the booth. Jamie looks appalled and moves over.

"Uh, wait a minute!" says Hudson. "Who are you again? We're very busy here this morning! I didn't get your name, either."

"Please excuse me. I won't take too much of your time," says the stranger, smiling warmly. "I represent a group of people who are just the biggest fans of your video game, CORPUS. The whole family enjoys it—the kids, the adults, who act like kids, the real adults…everyone. They enjoy it so much they just don't want to see it changed or come to an end, you know?"

Jamie and Hudson are looking at each other, expressionless. Hudson says, "Come to an end? I don't know—"

He is interrupted by the stranger. "You know, we just wouldn't want to see the game pulled from the market or altered in some way. You know what I mean."

"No," says Jamie, "we don't know what you mean! And who are you?"

"Please, please," says the gentleman, quietly. "Please keep your voice down. Who I am is not relevant here today. I represent a large group of your game players who feel they have a stake in continuing to play…a large stake…a very large stake. They know you are looking into the "mechanics" of your game, if you will, and would just really hate to see you change it … in any way… in ANY way! Do you understand me? Don't worry about anyone who feels your game has become too … "intrusive", for lack of a better word. That will take care of itself, you know."

Hudson starts to speak but is, once again, cut off by the stranger. "So, I'll leave you to your breakfast meeting, but let me add that those I represent love your game so much they just really want to show you their appreciation. Just a little token; that's all." With that, the stranger places an envelope under the table onto Jamie's lap. She briefly jumps a bit. "Again, I hope you both have a nice day and once again, do... not... worry. The wrinkles will iron themselves out." With that, the stranger takes a sip of his coffee, wipes his mouth with his napkin and says, "Good coffee, but certainly not Italian." He gets up and calmly leaves. The stranger's dress and warm manner of speaking and his suave demeanor, not to mention his hypnotically luminous blue eyes leave Jamie and Hudson, a bit stunned and at a loss.

Jamie and Hudson have expressions of befuddlement and fear. "Did that guy just put something on your lap?" asks Hudson.

"Yes," says Jamie. "An envelope."

"Oh my god!" replies Hudson. "What is this? Who is that guy? He's kind of, I dunno, scary? Where's Gerritt? Gerritt! Where did he go?"

Jamie replies that she doesn't know, but agrees the stranger acted like he was in the Mafia or something. "Jeez. People like this really exist? Are we being spoofed? Where's the camera? Also, I've never seen an Italian with eyes like that, have you?"

Hudson replies that he doesn't think they're going to find any cameras. "So, what's in the envelope?" says Hudson.

Jamie replies, "I don't know but it's pretty heavy. I'm really too nervous to open it. Do you think we should open it?? Jeez, who WAS that guy? I'm going to get Gary." Jamie leaves the booth and goes to the kitchen. The waitress and Gerritt are chatting in a low volume.

"Gary," says Jamie, "what was that? Who was that guy? He was kind of scary! Why did you bring him to our booth?"

Gerritt stares at Jamie in a manner she is not used to seeing from him. "I don't know who he was, but he made it very clear he wanted to meet with you guys. I told him I had no idea when you'd be in, having not seen you here for over a week, but he told me you WOULD be here, shortly, this morning. I asked how he knew that and if he was a friend of yours, but you probably noticed that he's a man of few words, and, yeah, he was pretty damn scary. I'm so sorry, guys. I literally didn't know what to do, but somehow didn't get the feeling that he was any threat to you. What did he want with you two?" Jamie thinks for a second and then says that she wasn't really sure and that he just wanted to talk about the video game they developed. "Maybe he wants to buy us out or something. Dunno. He really just didn't say anything much and then he left."

Jamie is concerned for herself and Hudson and Gerritt, too, if she discusses the meeting in too much detail. Jamie and Gerritt both agree the stranger was a weird and an ominous figure Then Jamie goes back to the booth with Hudson. "So?" says Hudson. Jamie tells him she didn't give Gary any details and also states that the stranger scared the crap out of him also. "Let's get out of here and go to the office and figure this out." They both leave.

Back in the office, they go straight to Hudson's office and close the door. Neither can decide if they should open the envelope. "What if there's money in it?" says Jamie. "How does this guy know we are looking into a problem with CORPUS? How did he know we'd be at Ruurlo's? We are being followed! WHY? Who are these people? Let's open the envelope… shouldn't we?"

Hudson stops her and tells her this is all so weird and they should sit on it for a few days. He takes the envelope and puts it into his office safe. "Let's give this some time to gel and get our thoughts together." Jamie agrees.

They carry on with their day, as normally as possible, and agree to, at least, revisit the issue in the morning. Jamie suggests they both stay at Hudson's parents' house tonight and Hudson quickly agrees. They arrive there and make a quick excuse for why they're both staying there for the night. Hud's parents, of course, don't mind at all. Jamie is like another one of their kids. All goes well and the next morning, they are off to the office. They go right to Hudson's office and sit down. They feel a bit refreshed. It's a new day… a fresh start. They agree to let the envelope sit for now. The phone rings and Jamie answers it for Hudson. While she is jabbering away with someone in the finance department, Hud is checking his phone. Jamie is looking at him while she is speaking. All of a sudden, Hudson looks up at her, stares, then makes a gesture to cut the call short and get off the phone, which Jamie does.

"What? What?" she says.

"Just saw a news clip from Reuter's that says that the Russian, deep pockets, really outspoken, climate-denying fool we were half joking about last Monday at lunch is now missing and his family and company are posting a huge reward for his safe return. The Russian government is investigating. The scary part is no one has requested a ransom… at all, for two days now! Oh man, oh man, oh man… I think this is part of the picture we are dealing with! I think someone grabbed this fool and probably did him in! Why else would no one claim responsibility and no one is requesting a ransom? I think we need to open that envelope."

They open the safe and take out the envelope. They both sit down, looking at each other. Finally, Jamie opens the envelope and takes out a thick pad of cash, holding it up for Hudson to stare at. "JEEEEZZZ!" says Hudson.

Jamie quickly starts counting it. "OMG! There's $200K here! Who carries around that kind of money? What do they want?!" Jamie starts to cry. "I don't want to be involved in anything like this. I don't wanna. This isn't me and I know it isn't you!"

Hudson comes over and puts his arms around her to comfort her, but he is also just staring into space, expressionless. He finally consoles her and sits facing her, knee to knee, holding her hands. "Okay, so let's call this what it appears to be. We're now somehow involved with some kind of organized crime. That strange guy at Ruurlo's may even be a Mafia operative. They're not just 'playing' our game; they're making money on it, a LOT of money, and they don't want us to 'fix' anything. We have two choices: we can notify the authorities and risk having a crime family after us… OR EVEN

AFTER OUR FAMILIES, or we can stop investigating the game and maybe have the authorities after us, down the road, and… and… and we don't even know who's playing the game! If a crime family is playing it, we wouldn't know it… or maybe we do now because of that guy. If someone with the police were playing it, we wouldn't know. If all this really dark stuff is happening because of CORPUS, we're not the ones doing it. It was never OUR intent. Maybe we should just put the money back in the envelope and back in the safe and not touch it. We could have plausible deniability that we knew anything about any of this."

Jamie agrees that they should just stop any intervention into the game's "misdirection" and carry on with business as usual, for a while. After all, what else could they do?

CHAPTER 11

Again... This Time with Plausible Deniability

Siberia is not all just great plains of snow and ice, which is a common western misconception. Novosibilsk, for example, is a thriving, beautiful city. It's a cultural and educational center in southern Russia. Its growth was stimulated largely by the development of the Trans-Siberian Railway, and it is now one of the larger cities in Russia. Several powerful oligarchs choose to live in Novosibilsk. Artur Vasiliev is (was) one.

Artur was a small-time coal mining operator from Moscow, having inher-ited the business from his father. He was a survivor when the Soviet Union broke up and his business was well funded by the Russian government and suspected crime syndicate money. He became a kingpin, seemingly overnight, with ties to upper levels of the government. His international forays into coal mining greatly expanded his business and his influence politically. He seemed to have no problem "convincing" smaller countries that his coal interests were globally important and that it would be in their best interests, financially, to do business with him. This level of influence did not go unnoticed by other governments, including the Chinese, the British and the Americans. It was thought, at one point, that he may actually be an operative with the FSB, a successor to the KGB; however, this could never be confirmed. All were fairly certain, however, that he was a member of a Russian crime syndicate.

Artur carried out his business dealings at home and abroad with increasing impunity. He fancied himself as a sort of king of Novosibilsk and was felt by many to have sociopathic tendencies. He was married to a beautiful Portuguese woman, but was frequently unfaithful. She did give him two sons, however—Maxim, the older of the two and Ivan, who was about five years younger. Maxim fashioned himself after his father, who was otherwise not all that interested in raising his children. His lessons were often harsh, yet Maxim was taking after him more and more. Ivan, on the other hand, took after his mother, a strong but compassionate woman who took an interest in raising him "right" after watching Maxim's development, under his father's tutelage. As the first-born son, Maxim is in line to inherit his father's business; especially since Cassandra, his mother has no interest whatsoever in the coal mining business.

On a snowy Sunday evening in Novosibilsk, Artur is out having brandy and cigars with his friends at a high end tavern. After sinking two or three brandies, the group breaks up to go home to dinner and their wives, in that order. Artur is wobbling to the front door of their home, after actually driving home. As he fumbles for his keys, he is grabbed by several men all dressed in winter white outfits. He is blindfolded and gagged and thrown into a very light gray paneled van and whisked away, without a sound. While struggling in the van, Artur gets his gag partially off and begins to scream at his abductors, who quickly hit him with a bat, knocking him uncon-scious. About two hours later, the van arrives at an old, no longer functioning mine and Artur is pulled out. A cell phone picture is taken of him. He is shot and dumped down an old shaft, previously used for bringing supplies into the mine. A small bulldozer then covers the shaft opening with rock and then snow and ice, sealing it. Artur had, however, first called home after leaving the tavern, to say he'd be there in about twenty minutes. Now, several hours later, his family is wondering where he is. Maxim makes a few cursory calls to his father's friends, as an investigation. This, of course, comes to no answer. Maxim then states that he will go out and look for him and barks orders that his mother and younger brother should notify the authorities. He gets into his car and just a few hundred yards from the home, his car is blocked, by a large truck. Maxim gets out and starts shouting, in his usual arrogant manner. He then realizes this may be a foolish move and he gets back into his car to quickly back up. Men quickly rush out of the blocking vehicle and shoot through the window, killing Maxim. His car is loaded onto a flatbed truck, with him in it. It is taken to the same abandoned mine where his father Artur was taken. There is no one there. He is dumped at the entrance to the mine and the gun used to murder him is placed in his bare hand, making it appear that he committed suicide.

Ivan and his mother are home now meeting with the police. They describe what they know and that Maxim went out to look for Artur. The police instruct Cassandra to call both Artur's and Max-im's cell phone right now in their presence. There is, of course, no answer. The detective of the local Politsiya conducts an investigation of the home and sizes up Ivan and his mother as being under very little suspicion. She instructs them to make themselves available at a moment's notice while the investigation proceeds. As the Polit-siya leave the home, Ivan and his mother direct a sustained glance at each other. Ivan puts his index finger against his closed lips. His mother seems to understand.

Artur had become as brutal a father and husband as he had been in his business dealings. This sociopathic pattern had been evidenced for years. Cassandra had been abused for any of her opinions, regarding the mining business. Ivan was considered a weakling by Artur because he did not appear to be following his father's footsteps, like Maxim. Maxim had actually become a carbon copy of his father, even though he hated his father's brutality towards him which was meant "to strengthen him." Maxim threatened his mother several times and got into fist fights with Ivan, ending with Ivan being hospitalized at one point.

After several weeks, Maxim's body is found and it is thought that, perhaps, he killed his father and then felt so much remorse that he took his own life. Artur is never found. When the local Politsiya detective return to the family home with the news, Ivan and his mother are "appropriately" and hysterically distraught. The news spreads quickly. The Politsiya are pressed to investigate further by the Russian government; however, a motive is lacking for any family involvement because neither Ivan nor his mother express any interest in the family mining business and it is passed to the Russian government, which gives a substantial, but really ridiculously low, sum to Cassandra and her youngest son. Therefore, case closed. Ivan and his mother shortly move to Paris and establish a better than modest lifestyle. Ivan… is a player.

Hudson and Jamie have, of course, read about the Vasiliev event in Russia. Their decision to carry on with business as usual is now being stressed further and is getting just a bit shaky. They had previously decided to take a "do nothing" approach to their game's technical issues, that they now realize may be being exploited by unscrupulous people in a very dark manner in order to gain huge monetary game rewards. They felt that they were not the ones harming people and the risk of publicly exposing the game's problems was greater than the risk of doing nothing and denying any knowledge of or involvement in sinister outcomes. Each news event, however, makes sticking to their decision that much more difficult. It was about to get worse.

Several days later in the office, Hudson and Jamie are trying to tie up any investigation or "review" of their game's performance, saying that the software and hardware had been reviewed and was working perfectly and that the company was doing very well financially, so well, in fact, that there would be a company-wide bonus to all employees as a result of their review and everyone's hard work. This, of course, went over smashingly well, brightening Jamie and Hudson's day … a bit.

Shortly after their company-wide announcement, Jamie was in her office working when her assistant came into the room. "Ms. Broadmoor, there's a woman in the lobby asking to meet with you. I think I recognize her. I think she's the woman you and Mr. Harthun met with once before … you know, the reporter or journalist or something? Her name is Sara."

"Oh! How irritating. Tell her I'm busy right now, and I'll contact her when I get a chance," says Jamie.

"Yeah … I already tried that," says her assistant. "She flat out refused to leave without seeing you. She told me that to my face. I was shocked. This one is aggressive. What's this about? Should I call security?" says Jamie's assistant.

"Oh, she's just looking for a good inside story on the world's most popular game and she's a pain in the butt," replies Jamie. "Okay, tell her I'm in between meetings and I can give her five minutes and only five minutes."

A few minutes later, Sara, the journalist, comes in and after some fake pleasantries, states she is going to write an article about their game, CORPUS, and just wanted to do Jamie and her business partner the courtesy of responding to it, prior to its publishing.

"What kind of article?" inquires Jamie. Sara tells her she has been following the major players in global climate news and feels that their game may, at the very least, be inducing a very aggressive kind of behavior among climate activists.

"That's liable," says Jamie, "and it's putting a very negative slant on our company and our product. We will sue you for liable and your cheesy tabloid, or whatever it is, for harassment and …we will … we WILL, make it stick! So, if that's what you came to tell me, we're finished here, and if you come back without an appointment or unannounced, we'll have you escorted by our security to the front door and held for trespassing until the police arrive."

"Hmm, interesting … me thinks the young lady doth protest too much. Interesting," says Sara.

"No, no protest, you're just a real pain in the ass, now please just leave … you irritating woman," says Jamie. Jamie's assistant now escorts Sara right out of the building and then returns to Jamie's office. "Don't ever let that crazy woman in the building again. She's the most upsetting woman I've met in years. Please notify security and tell them she is barred from the premises. Jeeeez."

"No problem," says her assistant. Jamie gets back to work, thinking she will discuss this with Hudson later in the day.

Sometime, just after lunch, which Jamie takes in her office, her assistant comes in again and tells her there is a gentleman in the hall requesting to speak with her. Jamie inquires who it is and she is told it's a Mr. Enzo Costa. "He said he knows you and knows you would be delighted to see him. He said he's like family."

"What??" says Jamie. "What's going on here today? Are we living in a nuthouse or at least the Twilight Zone? I don't know any Mr. Costa.

Please, please … just tell him I'm busy." As her assistant is starting to leave, Jamie stops her. "Wait, Lily … uh, what does this man look like?"

"Well, he's well dressed, dark hair … tall and thin, well mannered, friendly smile and has a really piercing stare with these luminous green eyes."

Jamie thinks for a moment, then says, "Huh … wait, this may be a distant cousin, Enzo. Haven't seen him since I was a little girl. I wonder … Yes, show him in, Lily, please."

Lily leaves to get the gentleman. Jamie gets on the phone to Hudson. "Hud, drop what you're doing. I don't care what it is. Get down here, right now!" Jamie doesn't wait for a response from Hudson. She just hangs up quickly.

On the other end of the phone, Hudson is stymied. "Jamie … Jamie … are you still there?" He hangs up the phone and rushes out of the office toward the elevator, his assistant trying to catch him.

"Mr. Harthun, where are you going? You have a meeting with the marketing staff in about two minutes, Mr. Harthun—" She is unable to catch him, as she sees the elevator doors closing.

Hudson gets off the elevator on Jamie's floor and moves quickly to her office. He knocks once and quickly enters to see a previous "acquaintance." It's the man from Ruurlo's. The man who never gave his name. The man who gave Jamie an envelope with $200K in it. The man who told them that the people he represents loved their game and didn't want to see it changed. The one who told them not to worry about any "issues" they may be concerned about. Hudson is stunned. "What the … Jamie, what's— "

He is interrupted by Jamie. "Hud, let me introduce Mr. Enzo Costa."

"Ahh … now we have a name. That's nice. Mr. Costa—" Hudson is interrupted again, by Mr. Costa this time.

"Please, Mr. Harthun, Hudson, if I may call you Hudson …

please, sit. Let's all be comfortable. I'm here to reassure you, not to make you nervous, really." Hudson sits very slowly. Enzo starts out, "We know you're both having a hard time with people in the media and probably also with your thoughts about what your game, CORPUS, may be contributing to. Please let me assure you, you have nothing to feel nervous or responsible for, really. You, yourselves, have done nothing wrong. Look at you both. It's not in you. There are people who get themselves involved in shady business. Then there're people like you, who are both as far from that sort of thing as anyone could imagine. Now, the fact that your game may be exploited by people who have secondary, financial interests, if you will, is not your fault nor your responsibility. Take it from me. I've seen it all. People will exploit anything they can. Yes, there are some people who are making a lot of money off your game. I represent some of them, but ... but, they are also doing really good things for this planet we all live on. So, you have a dilemma, right? The motive, for some of these people, is not "admirable," right? To make money. But the end result is good for everyone, isn't it? Including you and any children you may have—"

Jamie interrupts, "No, were not a couple; we're business partners."

"So you say," replies Enzo with a slight smile.

"So, you're saying the end justifies the means," says Hudson.

"Well, that's a phrase people like to use, in a derogatory way ... to mean ... it's a bad way to justify one's actions. It is, however, life ... it is. Right? Every day you see people in all walks of life doing things that are entirely self-serving and that may or may not harm others, and they don't seem to care. To say, well, I'm above all that, is, in a manner of speaking, to deny your humanity. Think about it. Yes, everyone 'should,' as individuals, try to do the 'right' thing, try to live a 'good' life. Other people define that for you and then you find they don't follow their own advice, so ... Nevertheless, my point is that you, with your game, have an opportunity to do the right thing for everyone, the whole human race … and, if the methods some people, not you, use to help you get there are 'less than virtuous,' well, that's not on you at all …at all."

Jamie and Hudson are, once again, speechless listening to Enzo. They wonder to themselves if this is sound logic or is he trying to just look out for the interests of those he states he represents. They're not sure. Enzo continues, "You know, I've grown quite fond of the two of you. I know I've only met you twice, but trust me, I know quite a bit about you both … and your families and your company and people you associate with and—"

"Wait! Stop right there. How do you know any of this?" interrupts Hudson.

"Ahh … have you ever heard the phrase: there are things that it's better you don't know the details about? Right out of an espionage movie, right? Ha ha. This is in that category. Believe me, you don't need to know these things, and it's better that you don't. Let me reassure you, you and your families have nothing to be concerned about. Let me get out of your hair and let you get on with your day." Enzo hands Jamie a card. "This is your way of contacting me. DON'T add this to your phone's contact list or scan it into your computer, etc. I know you young people do these things. It's not good, not good. Just memorize it, both of you." Enzo gets up to leave and takes out another envelope and hands it to Hudson. "No, this is not a "bribe". You see, I already know what you're thinking." He chuckles. "It's a gift. Consider it an early 'event' gift from our family to yours…. simple."

"What event?" asks Jamie.

Enzo looks at both of them and, again, chuckles. "Okay, okay … you young people, everything's a big secret. Okay. Enjoy your day." With that, Enzo gets up and lets himself out of the room, saying they need not worry about journalists and media people, including the one they just saw, and telling them that they are just self-serving and are really not there to help anyone. Hudson and Jamie sit down and just stare at each other. "What event?" says Jamie, as they continue staring into each other's eyes, with just a hint of a smile coming to both their faces.

CHAPTER 12

A Moral Dilemma…or…No?

In the ensuing weeks, Jamie and Hudson have many conversations about Mr. Costa's words of wisdom. It seems they are starting to conclude that he is correct. His "suggestion" that they have nothing to feel guilty about may be correct, and, at the very least, following their present course of "informed ignorance" seems to carry the least consequence for them and their families. They are starting to like Enzo, with his soft-spoken manner and fatherly advice. Nevertheless, they realize they don't really know him very well and don't know the people he "represents" at all! Still … they are starting to feel an odd sense of comfort with him and more settled than they have felt in quite a while. They decide to put Enzo's second "gift" envelope into the safe unopened since the money gifts still, of course, make them uneasy. They do follow some of Enzo's other advice. They stop watching the news too closely or regularly. They also have not heard from Sara, the journalist, for many weeks now. They don't fret about it much and don't talk about it … almost ever. It seems that a "calm" has come over both Jamie and Hudson. It's almost like they feel they have a guardian angel watching out for them. They are getting back to more "normal" life. They see friends again, go to dinner and shows, spend a little time with their parents and have planned a get-away vacation to Paris … together … just the two of them. Right. Only the two of them.

One evening, Jamie and Hudson both discuss a large family dinner with their respective families and brothers and sisters and kids—a night out because they have been so busy at work and there has been so much business stress they haven't been able to see and spend time with everyone and are feeling a bit disconnected. The parents agree, but it becomes a sophisticated undertaking, as it frequently is with extended families. Jamie's parents are all in. Her older sister Savannah, who is now a prominent fashion design figure, has to "discuss" it with her boyfriend of two years. She can clear her schedule but is uncertain about Robert, who is working on his masters in finance. Jamie's mother Rachel convinces Savannah to convince Robert.

Savannah briefly replies, "Why is this so important right now? Is something going on?"

Rachel replies, "No, nothing is going on, but your sister has had a lot of stress lately and it's a nice thing to get together with the two families."

Hudson's parents are also all in. His sister Jackie, who is now 31 and has had a permanent boyfriend of four years, according to her father, thinks it's a great idea to all have dinner and says that Adam can easily make it. Hudson's other sister Laurel, now 33 and married with two kids, is also all about a night out with her mother and father watching the kids in the restaurant. So, after much finagling, the combined family dinner is set. Both Jamie's and Hudson's mothers are very, VERY curious about why they are setting up a big get-together, having never done this before. They have secret suspicions that something may be going on between their son and daughter, but they don't even let their thoughts slip to their husbands.

It's late fall. Both families are getting dressed for dinner and looking forward to it. Hudson's older sister Laurel and her husband are getting the kids dressed and having a hell of a time doing it. "Oh, I hope mom and dad are up for taking care of these monsters for a good part of tonight!"

Her husband laughs. "They'll be fine. The kids act so much better around your parents than they do with us. Your parents are like a drug or something. It'll be fine."

Jamie has picked a nice Italian restaurant in downtown Denver, called Johnnies. It has subdued lighting, side rooms for 10 to 20 guests, a gorgeous old bar and well-dressed servers. "It'll be a night," she thinks. She has no idea.

Everyone arrives and gathers in the foyer of the restaurant. The hostess comes over and inquires whether all the guests have arrived. They have, and the hostess takes them to a small banquet room with beautiful Italian drapes and carpeting, marble columns and a long polished, mahogany dining table covered with a fine Italian lace cloth. Everyone is impressed.

Laurel leans over to her brother Hudson and says, "Wow, Hudson, you're paying for this one."

"Don't worry about it," says Hudson and he chuckles. When all are seated and the two kids are seated next to their grandparents and partially settled, Hudson stands and taps his glass. "Jamie and I suggested this dinner to say thanks to everyone for putting up with us over the last many months. Our business starting getting away from us for a bit. Lots of stress. Lots of stress. But, it's all calmed down now and things are running so much better. We've missed you guys and—"

One of the kids throws a dinner roll at the other, who screams and grabs for a roll, but Hudson's father prevents it and quiets them down. "Oh, thank god," says Laurel.

Hudson laughs and continues, "We've missed you guys and thought that instead of Jamie and me doing something like this with our separate families, we'd get everyone together … and, you know … let all hell break loose."

Everyone laughs heartily. Jamie stands next to Hudson and says, "Well, hopefully not ALL hell because we don't want to get thrown out … but we do hope everyone has a good time. Oh, and good things to try would be the Tuscan red wine and for those coffee lovers, after dinner, a decaf cappuccino with a shot of anisette is to die for. So…welcome."

They all settle in to conversation with those sitting closest to them. Jamie's older sister Savannah leans over to compliment her on how she's dressed. "Man, it's nice to see that I've had such an impact on your style when we were growing up." She laughs. "You could be taken for a model instead of a computer nerd."

Jamie laughs and says, "You know, I got this about six months ago in New York and haven't had a chance to even wear it yet, but I figured a big family dinner and all …"

Savannah counters with, "Yeah … what's with all this family stuff all of a sudden? Something I should know about, you know, before anyone else?"

"No … what?" says Jamie. "Don't be so suspicious. It is what it is. Just fun."

Jamie's mother catches up with her later that evening and inquires about the problems they were having in the company and what they ever did with the investigation into employees.

"Oh, that turned out to be a dead end. There really was no problem with our game. We met with a journalist wanting to do an article on our game, CORPUS, and then she accused us of creating a game that was bad for kids, etc. She really was just one of those paparazzi type people, who just want to make money by creating all sorts of lies. We just didn't speak with her and she's not a problem anymore. Case closed."

Jamie's mom seems pretty satisfied with and reassured by this explanation. Jamie then excuses herself to the ladies' room and her sister Savannah sees her and decides to join her. Once inside the ladies' room, Savannah begins, once again, prying about why they are having this fun, big combined family dinner.

Jamie responds, "Oh, jeez, Savannah, give it a rest. Can't you just have fun?"

"Yes, I can have fun, but you're my little sister and I want to know what's going on in your life. That's my job," She chuckles. "So, come across … what's up with you and Hudson? You seem…different, so family-oriented tonight. You were always so academic and business-like, but… not tonight."

"Well, I don't know. You know, I think it's just that we had such a stressful time with our game business. You know we considered shutting it down," replies Jamie.

"What?" says Savannah.

"Yeah, but that's all behind us, and I think we just needed to let loose tonight. We need to relax for a change. We need a vacation. We're going to go to Fiji for ten days next month and just drink fruity drinks and stay in those little cabanas right on the water. You know the ones…"

"With who?" says Savannah. There's a pause. "Just you and Hudson? AHA!"

"Oh, don't make a big thing… we just need to calm down," says Jamie.

"We… we…so much we, lately," replies Savannah.

"Oh, brother," says Jamie, "let's go back to the table."

Jamie and her sister are walking back to their banquet room when Jamie stops and looks into another room. Savannah inquires why. Jamie states that she thinks she knows the man sitting at the end of the table with his back toward them.

"You know him from the back of his head?" says Savannah. "I admit this man *is* distinctive, in his dress and mannerisms."

"Yeah," says Jamie slowly, "from the back of his head." And she proceeds to walk into the room.

"Where are you going!" whispers Savannah loudly. She abruptly joins Jamie.

"Mr. Costa, Enzo?" says Jamie.

The gentleman looks up at Jamie and then stands up, from the table. "Oh, my dear Jamie. So very nice to see you… and unexpected. Are you here with Hudson or family? And who is this beautiful young woman, you're with?"

Savannah is immediately stunned and flattered.

"This is my sister Savannah. Savannah, this is Mr. Costa… an… international game marketing consultant we've been…doing business with. He's helped us so much I can't tell you."

"Ah, thank you, Jamie. And I can see where you get your sense of impeccable style from." Enzo says, looking at Savannah, who is still in awe and struck by his smile and piercing eyes. Enzo goes on to introduce them both to his brother Luca and to his own wife, Valencia, his two sisters, Giulia and Giuseppina, and their husbands.

"So nice to meet all of you" say Jamie and Savannah.

"You know, we're just having a big family get together two rooms down the hall. I'd love it if you could stop by and meet my family and Hudson's family," adds Jamie.

"Really? A big family get together? I'm so happy to hear that, dear. Very happy for you. I'll be down shortly, before the dessert comes. You know, they have the best tiramisu here… well, the best outside of Sicily, anyway."

"Great," says Jamie and she and her sister excuse themselves.

"Holy cow!" says Savannah. "Who did you say that guy was? Man, if he was younger, I'd be throwing my underwear at him."

"Oh jeez, Savannah. TMI, TMI," retorts Jamie. "Get hold of yourself."

About ten minutes later in the Harthun Broadmoor banquet room, Mr. Costa enters with his two sisters and his brother and sister-in-law. Jamie and Hudson both pop up and go over to them.

"Oh, we're glad you had time to stop in and see our families, Mr. Costa," says Jamie.

"Yes, absolutely," says Hudson. Jamie's father and then Hudson's get up and thank the Costas for their help with their kids and for bringing their families together this evening.

Mr. Costa replies, "Think nothing of it. It's our pleasure to help out your young people and especially your families. Family is everything to us, too. You're nothing without it and we're delighted to have met you all, really." With that, the Costas excuse themselves and walk out by Jamie and Hudson. "Family really is everything, you know. We'd like to someday consider you both as family also," whispers Enzo.

Jamie and Hudson are taken aback. "Well, when we both start families, we'd like that also," says Jamie with Hudson agreeing.

Enzo looks at both of them and just smiles. "Yes, of course, of course… when you both start families," he says with a broad smile and another long chuckle.

"Pay him no mind, you two," says Enzo's sister-in-law. "He thinks he knows more than he does. All men think they know… but they know nothing," she says with a smile and a glance at both Jamie and Hudson.

"Both of you go back and enjoy the family dinner… very important for you at this time," says Enzo's sisters and they and Enzo depart.

After they're gone, Hudson looks at Jamie and says, "Are they talking about what I think they're talking about?"

"I think so," says Jamie.

"Ah," says Hudson.

Jamie looks at him, with a pleased look in her eye. "Would it be…so bad? I mean I'm very happy lately, aren't you? In fact, I haven't been this happy in a long…I can't even remember when. Aren't you happy also? I mean, with the business and…(long nervous pause) with me?"

Hudson looks like someone just punched him in the nose, but he quickly regains his composure. "Well… (very VERY long pause) yes, actually. Yes, I am. Yes, and with the business too." Jamie smiles, takes his arm and they walk back into their banquet room and sit across from each other.

Hudson's mother asks where they met Mr. Costa and Hudson says he doesn't really remember. Jamie's sister Savannah chimes in and says, "Well, all I can say is, if that guy wasn't married, I know a few ladies at work who would just love to meet him! Hey, are they the people who have you guys all into family now? I mean, I guess I understand that. He's like a good-looking devil from a mob family, right? The bad boy, now all grown up." She chuckles.

"No," says Jamie, "in fact we just met his wife tonight, like you did."

The dinner conversation continues, and the table is buzzing about many things until Savannah says, "Yeah, and these guys need to relax. So, did you know they're going to Fiji next month?" A pin drop, in the room, and everyone hears it. All eyes are now upon Hudson and Jamie.

Hudson's older sister Laurel breaks the silence. "What? Wait, who are you going with? (pause) You two, just you two?"

"Aha," says Savannah. "That's just what I said!"

Hudson says, "What's with you people? Can't we even get away for a few days to decompress without it being a big thing?"

"No," says Laurel, "no, that's not how it works at all, my little brother. We all want the details. First, you're all about this 'family' stuff. Now you're going on a romantic vacation, together and…"

Hudson immediately looks uncomfortable.

Jamie's father chimes in, "Stop, stop grilling them. They just need a break. Is that okay with everyone? If Jamie was going with one of her girlfriends, you wouldn't even give it a thought. Seems kind of sexist to me." He laughs.

"Sexist, chuckles Savannah. Dad, sit down."

Hudson's mother agrees, "Just let them relax and let's just have a fun family night."

Laurel looks at her mother and says, "Mom! Who are you?"

The conversation degenerates but the subject slowly dissipates. A really fun and memorable evening is had by all.

CHAPTER 13

Reflection

It was a balmy day, again. Hudson and Jamie were lying in lounge chairs on the porch, outside their hut in Bora Bora, French Polynesia. Full of fruity drinks and looking out over the expanse of crystal blue ocean with only the slightest breeze upon them, they were a postcard of island relaxation. After days of chit-chat about the sky, the water, the fish and the drinks, the conversation changes.

"Man, that conversation at our family night…right?" says Hudson.

"What do you mean?" asks Jamie. "I think we covered everything from work to politics to your sister's kids' school clothes."

"Not what I mean," says Hudson. "You know, the part about… (pause) us, as in **US**."

"Ahh. Well, we've been friends for a long time, our whole lives, basically. No one knows me better than you, and I feel I know you that well also. I mean I have girl and guy friends, mostly at work, but…you're my best friend."

"Yeah," replies Hudson. "Yeah, I feel like that, too, and I've been asking myself over the past few days….why ARE we here together?

I mean…I asked myself, lately, what my life would look like and feel like if you were not part of it…ya know? and…I didn't like the feeling. You asked me at the family dinner when Enzo was, once again, implying there was something between us, … you asked me if that would be so bad. That question was a turning point for me. I started thinking about you … in a whole different light."

"That's romantic," says Jamie.

"Well, romantic or… you could call it… uh—" says Hudson.

Jamie interrupts him, "Romantic, Hud. It's romantic…that's the word." Hudson looks at her. "Kiss me," says Jamie. They pull their beach chairs together. Hudson leans over and gives Jamie a quick kiss on the lips, and then smiles. "No…," says Jamie…and she leans in and kisses Hudson long and hard. They separate briefly, staring at each other for several seconds and then Hudson moves over to Jamie's chair, leans against her and kisses her like a teenager would kiss his girlfriend after not seeing her over the weekend.

Hudson looks at Jamie and says, "No, to answer your question, again…no, I don't think this would be so bad." And he kisses her again. After a few minutes, they decide to take it back inside their cabana hut, where they waste no time getting into bed.

Hudson then pauses to ask Jamie a biology question and she stops him. "It's okay. I'm prepared. It's not a problem."

Sometime, several hours later, they wake up, when the sun is just starting to set. "Well," says Hudson, "this certainly changes things. Looks like everyone else had this all figured out and we were the last ones to know." He laughs long.

"You're my best friend, Hud. I love you…I mean it…I love you, Mr Harthun."

Hudson smiles and looks into Jamie's eyes. "I love you, too, Ms. Broadmoor, and yes, you are my best friend, too. I really don't know what I would do if you were not in my life."

Then Jamie starts to giggle. "How on earth are we going to tell people about this? People are going to be like, 'What?'"

Hudson laughs and stares into Jamie's eyes "Jamie...(pause) I don't think it's going to be hard at all. THEY already know; we are the last. Although, some guys may wonder what a guy like me is doing with a model like you." He chuckles.

Jamie replies, "It's just the clothes, sweety."

Hudson quickly corrects her, "It's not the clothes, Jamie, it's not the clothes...it's you." They stare at each other, longingly and then decide to forego dinner.

Before their vacation get-away is over, Hudson is browsing some shops, looking for some better snorkeling goggles. As he passes by a jewelry store, a ring in the window catches his eye. Never one to give jewelry a second or even a first thought, he stares at it. You can almost see the gears turning as he mentally runs through a program of what his life with Jamie would be like. He snaps out of it and wastes no time going in and making the purchase. Hudson is delighted with himself when the sales person compliments him on his truly exquisite taste.

"Well, this has been a long time coming. I wish I had not wasted so much time."

"I'm sure your time has not been wasted," says the saleswoman, "and there is no right or wrong time...only the time when you are ready. That is the perfect time." Hudson feels much better.

Hudson and Jamie are back together at the hut. They are departing tomorrow. Hudson feels that tonight is the right time to ask Jamie to marry him. They are having a really nice dinner delivered to their hut, with drinks and wine. Hudson says it's because they need to pack and a nice dinner will commemorate their time in the islands. After gathering some of their luggage, they sit down to enjoy their meal and wine, after which a fancy flaming dessert is served.

Hudson starts, "You know, Jamie, I have given this a lot of thought over the last few days. You're right. I've been happier the last few months than I've been in a very long time. It's not just that the pressure is off, business-wise now. It's very clear to me now that the reason is… you. I feel that I've always been able to work with you, rely on you, work through things with you, and it's made me very comfortable. I love you, Jamie. I've loved you for a very long time and I guess I just needed a good slap to wake me up to it. I simply can't even imagine my life without you at all." He gets down on one knee, shows Jamie the ring and proposes to her, while the servers are still delivering dessert. They are struck by Hudson's proposal and immediately start taking a video of it on their phones to give to the couple before they leave. Jamie is beside herself and starts to cry and smile. She, of course, accepts, and the servers start to clap and cheer. They start to leave and say they will come back after a while to clean up the dinner plates.

As they leave, Jamie and Hudson fall on some of the clothes they laid out to pack. They forego dessert… and not because of the calories.

Jamie and Hudson travel home the following day. They each see their parents and tell them they are very refreshed when the parents ask if they had a good time. Hudson's parents seem to buy this. Jamie's parents, however, (especially her mother), do not, and her mother looks at her for a few seconds and breaks out in a huge smile. "Do you have something to tell me, dear?"

Jamie knows there's no point in trying to hide her elation and throws her left hand out in front of her mother, who immediately screams and hugs Jamie and yells for Jamie's father, who comes running as if someone were having a heart attack. "Our baby is getting married!" exclaims Jamie's mom.

Her father is delighted and then says, "To who?" Jamie and her mother look at him and roll their eyes. The father laughs. "Jamie, your mom and I love you very much and love Hudson and we couldn't be happier for you both. Do his parents know yet?"

"No, we were going to keep it a secret for some big announcement, but when mom looked at me, she just knew. I couldn't contain myself. I'm going to call Hudson now and tell him I spilled the beans already and he should tell his parents. Also, when Savannah and Hudson's sisters find out, they'll go crazy. We should all get together, maybe here, for drinks and desserts and a celebration, right mom?" Jamie's mom agrees.

CHAPTER 14

The Proverbial Roller Coaster

Life is funny, until it's not. One minute you are riding high without a care and then, wham! Out of nowhere your life is upside down and you never saw, nor could have seen, it coming. The secret is in being able to adapt. Stress is psychophysiological, part emotional and part chemical. It's also part of the human condition. You can scream at the door all day, and the door feels nothing. No stress reaction at all. People are, of course, different. They have emotion and reactivity due to stress chemicals. You can let stress flatten you or you can look for "work-arounds," ways of quickly adjusting and carrying on. Sometimes work-arounds present themselves. You just need to be aware of their presence and be willing to accept the method.

Hudson starts looking at the news again, now that they're home and back at work. An article about countries who profess to be "green," while making a lot of money off enterprises that are detrimental to our very survival, catches his eye. While Norway is considered by many to be a very progressive, environmentally conscious nation, the country relies heavily on off-shore oil and gas production;

it's about 20% of their national revenue. A journalist in Denmark has been following Scandinavian governments that are heavily involved in such activities. He notes in his article that not only have they recently had to deal with a moderate size oil spill from an off-shore drilling platform, but they also have plans to open at least one new platform to increase production. The journalist notes that the government has just pledged a large amount to combat global warming. He notes that this seems to be the standard operating protocol now in many well-to-do nations; destroy the planet with one hand and then pledge money to fix the destruction with the other hand to keep them in a positive public relations balance. Nothing else was noted in the article, which was about a week old. Nothing about protests, ground swell opposition, boycotts; nothing. The journalist did coin the phrase, "EACH" though, for Environmental And Climate Hypocrites, and concluded that they are as bad or worse than those fossil fuel producers who continue business as usual and apologize for nothing and make no pledges to remedy matters; you know, those who are 100% self-serving and don't care if people know it.

Hudson goes about briefly researching this and subsequent articles, to see if there have been any repercussions. Nothing shows up. He briefly has a sigh of relief and then begins to wonder if that is the proper emotional response. He considers whether there would be any productive changes in this corporate M.O. without direct and perhaps aggressive intervention. He quickly feels bad for even thinking like that, but plans to keep an eye on this event to see if anything ever comes of it. This doesn't take long at all. Within about ten days, Hudson is once again horrified. This time, innocent people are collaterally killed.

Platform NC-11 is one of the flagship platforms designed by the Norge Oil Subsidiary, NOS. As one of their newest and most productive platform designs, NOS is very proud of it and uses it, periodically, in PR ads and ads for investors. An NOS executive VP is scheduled to spend two days on the rig, filming and having photo ops and marketing meetings. Several other company officials are with them; however, he is the "big show," except for a government regulatory official, who will also attend as usual. After the first full day, the guests are treated to a wonderful dinner, looking out over the sea on a very calm night. Later, they all retire after planning their second day. At about 3 am, there is a repetitive noise that the VP notices in an environment which is generally always noisy. He gets up to look out the left window of his corner room at the helipad to see if someone is coming aboard the platform. Seeing nothing, he heads back to bed, but takes a quick glance out the other window, just looking out to sea. He is frozen, in the last glance he will ever have, just long enough to see a helicopter, hovering and launching a rocket at his corner room. The rocket lays an explosive and fiery waste to his room and three or four others next to his. The NOS VP and government official and, unfortunately, several other peripheral NOS and journalistic people are also killed.

The chopper flew in very low, very quickly and did not respond to any tower identification requests, which are customary. It took off in the night as quickly. On shore rescue personnel are immediately mobilized as is the Norwegian Air Force. The military finds nothing and the usual investigation is launched. There are no paramilitary groups taking responsibility for the attack. No ransom is requested. There is, however, a small reward offered for information leading to the arrest of those responsible. Shortly thereafter, a non-digital warning is found scattered on several streets in larger cities by an unknown climate group. It warns of additional lethal interventions against those who continue to destroy the planet. These are immediately researched, which... of course comes up empty in this internet age.

Hudson is, once again, shocked. Is this related to their game? Should he tell Jamie, if she doesn't already know? He thinks not. She is in one of the happiest places of her life right now, planning a wedding, and he doesn't know for certain whether this has anything to do with them. Actually, he thinks back to Enzo's advice; it definitely has nothing to do with… "them." Thinking about this advice, he even starts to feel better and is able to move on with other important events in his life right now.

Several days later, Jamie gets a call. Her assistant relays to her that an old college friend is on the line, if she has a moment. Jamie asks who it is and her assistant tells her the person wouldn't say, but says she knows of the event. Jamie quickly assumes it's about her wedding plans and can't wait to find out who it is.

She picks up the call, "Oh, hi. I just wanted to let you know I know about the recent event, too, and—"

Jamie stops her and says, "Wow! That was fast. Did you hear this from my mom or sister and I don't recognize your voice. Is this Becka?"

"I don't know any Becka, but you do know me. This is Sara, the journalist you had thrown out of your office. Remember me? Well… I'm baaack, and I don't know your mother or sister, but I suspect we're talking about different 'events' here. I just wanted to let you know that your watchdog is not getting rid of me that easily. I believe very strongly that your company and your video game have a direct relation to that oil platform bombing in Norway, and I'm going to expose it. Just wanted to know if you wanted to respond to the issues in advance. I'm giving you that chance."

Jamie feels like she just got hit in the head with a brick. She has not paid any attention to the news for a while and has no idea what the Norway bombing is even about. She just hangs the phone up and sits quietly for a few minutes. She then asks her assistant to find her fiancé and get him to come to her office. Ten minutes later, Hudson shows up and, says, "Hey. What's up? … pause…wait, what's wrong? You look like you've seen a ghost."

Jamie replays the story for Hudson, who responds, "ohhh, noooo." He describes what he knows about the Norway issue. He then tells Jamie how he wishes they could, at least, touch base with Enzo. Barring that, he tells her how he has dealt with the news over the last few days and how he feels better after remembering Enzo's advice. They both decide to take an early lunch. It doesn't take long for Hudson's wish to be granted.

After a long lunch with some heavy conversation, Jamie and Hudson both feel a bit better and return to the office. Hudson walks Jamie to her office and then he proceeds to his own, only to see Enzo sitting just outside his door, waiting. "Enzo! I was just telling my fiancée, I wished I could speak with you for just a bit and here you are. Come in, come in."

Enzo replies, "Your fiancée!? Just kidding … I already knew that. That's the best news. I'm so happy for you both, but, of course, that's not why I'm here, and I suspect you know that already." Hudson states that he thinks he may know why Enzo is present. "Yes, that's about right," says Enzo.

"The journalist… the one that has been… let's say, a problem for you and, therefore, for us is very tenacious. Previously, we tried to scare her with the loss of her job, tax audits, jail time…just all sorts of things. We tried to reason with her; really, we did. Now we see that none of that has worked. I just wanted to reassure you that now it will be handled… differently and you should not worry any further about this."

"What do you mean, handled?" Hudson says slowly.

"Hudson! This is not important. You don't need to be concerned… at all. I would suggest, again, that you always keep in mind that you and Jamie are doing good things. I, myself, and the people I represent, understand you and your fiancée. You are…. good people; you are compassionate people. You try to do right by everyone. Your employees love you. That there are people in the world who will twist what you do, will exploit it, will try to bring you down… that has always been… and will always be. Pay it no mind. You two can and should stay 'above' all that. Do you understand what I'm trying to tell you?"

"You know, Enzo… I think I do. Jamie and I have gotten quite fond of you… and your wife, Valencia. When we first met you, to be honest, we were scared of you, but, over time, we realized we were stuck between a rock and a hard place. It was a real dilemma for us. We finally realized we had to pick a side and the side we felt was there for us, that offered sound advice, that offered us a sense of calm again, was you. I do understand what you're trying to tell me and that you're sparing me the details. Whether that's because of your own interests or for ours doesn't matter now. I get it and, what's more, I appreciate it. So does Jamie."

"Good," says Enzo. "I knew you would finally understand. You know… it's just life. We all like to pretend that it's not like that, but it is. It's… the good and the bad, the yin and the yang; like that. So, fresh page now, what are your wedding plans? Did you get her a nice ring? You know, Valencia… she will want to know these things."

Hudson and Enzo speak for another twenty minutes or so. It's like they are old friends and Hudson seems very comfortable with Enzo now, so much so that he invites him and his wife to the wedding and asks where he can send the formal invitation.

Enzo appears delighted, "Oh, that's wonderful. Thank you so much. Valencia will be delighted. Please, please, if you could just hand Valencia a formal invitation at the wedding, she will want to keep it. We'll find our way there. Just plan on it."

"You know, Enzo, I am sure you will," chuckles Hudson. With that, Hudson and Enzo bid each other good day and Enzo leaves, stopping by Jamie's office on the way out. He stops at her assistant's desk.

"Oh, Mr. Costa," says Jamie's executive assistant. "Did you want to see Ms. Broadmoor, or me, perhaps?" She laughs.

"No, dear," replies Enzo. "Just wanted to find out how our girl is doing."

"Oh, she is okay right now. You know she is a really sensitive person, but she's doing fine. I assume you saw Mr. Harthun, then. Did he invite you to the wedding?"

"Yes, so we will be seeing you there, just so you're not surprised, dear. Also, I wanted to tell you what a great job you're doing. I know you'll keep me advised of the overall status of these two."

"Yes, as always, Mr. Costa."

"Thank you, Gianna … or Lily is it now? See you soon," replies Enzo, and he leaves.

Sara Meclin is now working feverishly to finish her story about the Harthun—Broadmoor Entertainment Company and Hudson and Jamie, in particular. She is just about finished, but decides to leave the finishing touches for tomorrow. She packs up her belongings and leaves the office. As she is walking across the street to her car, she is suddenly walking very closely, in the crowd next to Mr. Costa. She shoots him an angry look.

"Hi, Sara," says Enzo.

"You know there are cameras everywhere out here, and you don't scare me at all. I don't care about you threatening my job or getting the IRS after me. Just who the hell do you think you are?" Sara says as they arrive at her car.

"It's clear to me that I didn't make my point previously when we met," says Enzo. "I'm going to be very clear this time. I work for a group that has an interest in the people and the company you are writing your article about, and I don't care, by the way, about any cameras here in the parking lot; it makes no difference to me. I am now making you an offer, Sara. We are not without compassion, and I would not make this offer to many other people. I do know what's going on at your magazine. I know you are at risk of losing your job and you are hoping this article is going to save it for you. I know, more importantly, that you are a good person; you do or did have scruples and you work for your 'truthful' magazine because you want to feel you are doing the right thing and you want to feel, perhaps, that you're above it all. I just want to tell you that we understand and admire that, but you are NOT doing the right thing here."

"The two people you are after are good people. They have no malice in their hearts at all. Their video game is designed for education and to help protect the planet we all live on. If others decide to exploit it, for their own gain, that's NOT your concern and certainly it has nothing to do with the two young people who designed it. You're doing the WRONG thing here. You may win a personal battle, but lose the war, for all of us. So…here's my final offer… and make no mistake; if you decline this offer, well … your brother and his family will disappear forever. Your magazine will eventually cease to exist. Your elderly parents will unexpectedly pass away of unknown causes. I'm sure you're not finding this a pretty picture."

"Are you… hearing me? I need to know that you are hearing me and be absolutely clear that you need to make the 'right' decision here. Get rid of your article. Do NOT publish it. Do NOT give it to anyone. Quit your company. Tell them you have an offer in Europe that is a dream job for you. My organization will set you up in southern Italy. A really nice place to live, a career in writing… say…fiction, this time. A nice guaranteed income. A nice life. Yes, you can visit your family in the States, periodically. Understand, though, that we will watch you. You will have friends there, but we will watch you. The alternative is, I'm betting, a very unpleasant thought for you. I am, once again, going to gamble on you, which I rarely do. I understand people. Don't prove me wrong. Oh, and if you think calling the police will make this all go away... Sara... don't bet on it...and don't bet your families' lives on it. It's a very bad bet, Sara. Now... go home, think this over and I'll call you at home in the morning. Don't go to the office. Call in and tell them you are not feeling well. I'll meet you later in the morning for coffee. You know, Sara, don't think of this as a threat. Consider it an opportunity."

With that, Enzo walks away. Sara just stands by her car, watching him leave and, this time, shaking. She gets into her car and drives home, almost getting into an accident, for what will prove to be a very, VERY long night, consumed by thought and moral dilemma. She carefully considers Mr. Costa's offer, but can't get past the threat to her family.

CHAPTER 15

The "Family" Wedding

Valencia Costa is home in Sicily, taking care of some "family" business. The mail is delivered to their country estate by their estate manager, who mentions a particularly beautiful envelope.

"From Mr. Costa, in the States, do doubt." He smiles.

Valencia takes that envelope and puts all the other mail aside. She opens it to find an invitation to Jamie and Hudson's wedding in Denver, Colorado. The message inside is so gracious, it brings a tear to Valencia's eyes. Now, Valentina can be a sensitive and compassionate woman, but she can also be, if necessary, an ice-cold, all business, dictatorial personality. A very pretty, fashionable, Italian woman in her early 50s, she manages "family business" in a sometimes autocratic manner, but not always. Yes, family business includes gambling, loaning money, financial interests in high-end properties and many other endeavors, but never prostitution. Her father was a very strait-laced, but gentlemanly person. He was strong and respected by everyone he met, pretty much like Valencia's husband, Enzo, who also has an "air" about him that commands respect and attention, without even saying a word. He's a man you would not casually meet once and then forget immediately.

Valencia is delighted with the invitation and sits down staring at the card. She starts to cry. Valencia and Enzo had a daughter, Gabriella. She was an only child and the absolute delight of both her parents, but especially Valencia, who, like most mothers, had such high hopes for her daughter's future. She would study at the best of schools, meet a kind and strong young man, have a huge family wedding and at least three or four children. Sadly, at age 15, Gabriella developed an acute leukemia and was not responsive to treatments, which she received all over the world. Just before her 16th birthday, she succumbed to her illness. While her father was distraught for many months, her mother was inconsolable and she plunged into a deep depression from which she would not recover for almost three years. All her hopes and dreams vanished, like a vapor in the hot summer sun.

She retreated to their Tuscan villa for well over a year, where she would just while away the hours, reading and strolling and looking at pictures of her daughter. Enzo would spend most of his time with her, except when he would have to leave to personally take care of some family business. These trips were brief and upon his return, he would find his wife, still in the state he left her. Enzo had no idea what to do and he feared that he may lose his wife, also. She had seen psychiatrists, tried medication, tried some social activities, but she was not a believer in such things. She had lost some part of herself that she could not ever replace. Sometimes… this is not the case, however. People can be surprisingly resilient and take a new path in their road of life, when one least expects it. This was not the case with Valencia.

Now, reminiscing about his original trip to the states on business, Enzo realized he needed to pay a visit to two young people who have designed a video game that is educational for children, but also has a monetary side to it, and has quickly become a significant part of the "family" income. He had learned that there was a risk of this changing very soon, and after speaking with his consigliere and several younger but adult members of the family, he traveled to Denver, Colorado to "meet" the two game designers. Enzo was immediately surprised by Jamie and Hudson, but especially Jamie. Her warmth and mannerisms reminded him immediately of his daughter. Enzo was feeling something he has not felt in several years and that he never expected to feel again. After his meeting with Jamie and Hudson, he felt he must return to Tuscany and tell Valencia of these two "children" he met. He had a photographer, whom he worked with on several occasions, secretly take a few good photos of Jamie and Hudson for him to show Valencia.

The photographer was surprised, later delivering the photos to Enzo and exclaiming, "They're alive."

"Yes… very," replied Enzo. He brings the photos back home to Italy and showed them to Valencia, while he described them, especially Jamie. Valencia was overcome, staring at Jamie. Enzo saw a spark of life in her eyes he has not seen in years…and a tear. She wanted to know as much as she could about both of them, telling him to go back and "hover" over them for a while and learn what he can about their game business and any threats to it for the "family," but Enzo knows… it is not just for "the family." Valencia kept the pictures in the same book where she has kept Gabriella's photographs and awaited word from Enzo.

Now, Enzo has recovered from his trip down memory lane and in the afternoon and at Enzo's earlier request, Valencia receives a personal call and invitation from Jamie to the wedding. Valencia is momentarily stunned to hear the woman on the other end of the phone, personally inviting her. She has imagined what this young woman must really be like for a long time. After just a few moments, she starts to feel oddly comfortable with Jamie. The style and tone of her speech are so familiar. She can picture both Jamie and her daughter in her head, as they talk.

"Dear, my husband Enzo has told me so much about you and your fiancé, Hudson. I feel that I know you and can't tell you how much I appreciate you personally calling to invite me to your wedding. I just received your invitation in the mail this morning and was so delighted! Wonderful invitation! Did you and your mother pick that out? I know your fiancé had absolutely nothing to do with it. When these things need doing, men all run, like little boys in the streets of Pamplona."

"Yes." Jamie laughs out loud at Valencia's remarks. "Yes, my mother and my older sister helped and yes, yes, Hudson and my father all suddenly had things to attend to and they all ran!"

The two speak for what must have been an hour and a half, talking about each other's lives and families and even about Valencia's daughter, who passed away several years ago. They both sense an odd calm and openness with each other, like they have known each other for many years. Jamie feels a sense of security from this woman that she has not recently felt. Valencia feels that she has made a connection with Jamie that will move her life forward once again and offer a sense of closure after her daughter's passing. She will never forget her daughter, but she now has something or, perhaps, many things to look forward to with Jamie.

After the phone conversation, Valencia immediately calls Enzo and tells him she believes she has gotten a second chance at life. They speak for twenty to thirty minutes, and Valencia relates all the things she and Jamie discussed and all the hopes she has for their future involvement in Jamie's life. If she could have actually seen Enzo, at that moment, she would have witnessed the world's greatest and longest sigh of relief. The small staff at the Costa's Tuscan villa, notices an elevated level of energy and a smile on Valencia's face for the first time in years. The staff chatter and busy-ness is palpable. Life itself has returned to the Costa's villa.

Home in Denver, wedding plans are coming together. Rather than have a trendy "destination" wedding, in Honolulu, Havana or Helsinki, Jamie and Hudson decide to have a "hometown" wedding, much easier for most people to get to and much easier to plan… although not "easy." Their wedding is to be large. The two have made friends all over the world and both were born and raised in Denver, and they have many friends there, as do both sets of parents and siblings. They also want to invite "their office staff," having always been very supportive of both of them. The number of guests comes to just about 300. They find just the venue for both the wedding and the reception. A large, stately old mansion on a rolling green lawn, The Grant Humphrey Mansion. This is an exquisite, massive, historic property with the elegance of a bygone era. In good weather, they can have an outdoor ceremony, people can stroll the lawns, and they can peruse all the rooms in the mansion. There will also be many small open bars and the photography / videography potential is unlimited. This will be their venue.

Invitations are sent out. The maid of honor, Jamie's sister Savannah, is notified and thrilled as a fashion designer to be able to help decorate her mother, her sister and the bridesmaids. Hudson's best man, his brother-in-law (Laurel's husband) Robert, is also very happy to be involved and set up a bachelor party, having received deadly serious instructions from Laurel on what "type" of party this is to be and what type it is NOT. Now, Robert is an "adventurous" guy. He is always pushing himself to do things he has never done before—sky diving, surfing, motocross, Rugby, free diving, mountain biking, etc. You know, an adrenalin junkie. If he can't have the typical "stripper" party, he is going to do something else exciting. He decides that since they live in Denver and most all the guys are very accomplished skiers, the bachelor party should be held in Vail and the next day, they can get some instruction and all try kite skiing.

The party in Vail Village the night before goes just great. Hudson and all the groomsmen have a terrific time; they have too much to drink and the next morning, they are all feeling "just swell." After about 90 minutes of instruction, they are all fitted with small kites in an open field just outside the ski area. Each is given some last minute instruction and then they give it a shot. Some never leave the ground. Some get a few feet in the air, freak out and just fool around the rest of the morning. As things sometimes happen, a small gust comes up just as Hudson leaves the ground. Getting about thirty feet in the air, he gets a bit confused and his lines get twisted. As he tries feverishly to untwist them, while the groomsmen and his best man watch, with his throat in "oh my god" position, the lines never untwist. Fortunately… very fortunately, the kite is still catching enough air to bring Hudson down, albeit with a pretty hard landing.

Robert and the other groomsmen and the instructors rush over to check on him. He is fine, landing, of course, in the snow, but there is, nevertheless, a massive sigh of relief in unison. Robert is freaking out thinking about his wife Laurel finding out about this. "OMG! My life would be over. Okay… everyone is sworn to secrecy here! No one breathes a word of this."

The groomsmen say, "Don't worry about it, man. He's still alive and probably still functional." Hudson and his friends all laugh.

Jamie's sister Savannah plans the bachelorette party, to the last detail. They all go to Vegas, stay at the Bellagio, have a big dinner, drink to excess and have private gambling with shirtless bartenders and dealers. Then, they hit the strip, drink some more and gamble a lot, especially Savannah, who has to be wheeled back to her room on a hotel luggage cart. Everyone takes cell phone pics. Sleeping late, the next morning is mandatory.

Two days later, the wedding, itself goes off without a hitch. The bride, groom and all of the wedding party are in fine shape. Jamie's sister Savannah and their mother have Jamie dress in a gown to shame many designer fashion shows. Hudson is dressed "to the nines."

"On this occasion, Hudson does not look homeless," say two groomsmen, laughing.

Both sides of the aisle are filled with family and friends, including both sets of parents and right behind them are Enzo and Valencia Costa, occasionally chatting with the parents. Valencia appears fully recovered from her depression and is energetic and impeccably dressed. Enzo is his usual dapper self, both in dress and affect. His piercing blue eyes, from his Swedish mother, and demeanor always cut a striking figure.

The newly married couple's families are finally given a chance to get to know the Costas at the reception. They are seated at the parent's table, along with other people their age. This is by design, of course. Both Jamie and Hudson want their parents to see the significance of the Costas in their lives. The sisters and significant others and some other younger people are seated at a table right next to the parents.

After some icebreakers and assorted small talk at the parents' table, Jamie's father, seated next to Enzo, starts to inquire about business. "So, Enzo, you are like an international marketing manager for the entertainment company. Correct?"

"'Well, you could say that I have many connections around the world, especially in Europe, and I am able to put people, like your daughter and now her husband, in touch with people who can do them a lot of good. So, I suppose you could say I'm more of a PR guy," replies Enzo. "We also watch the companies we deal with very carefully, to make sure the connections we establish are… mutually beneficial."

"Wow," replies Jamie's father Alejandro. "I don't think I've ever heard of such a comprehensive service. They must be paying you a fortune."

"Not as much as you may think, actually. After initially helping them with some… pretty big obstacles, my wife and I… Oh, my wife, Valencia, is in the business with me… took a liking to both your daughter and her new husband and wanted to make sure they… got off on the right foot, you could say. It's good for our business, in general."

Hudson's father leans in to the conversation, saying, "Well, we want both you and Valencia to know how much we appreciate your help. They went through some really rough business patches and now, perhaps, thanks to you, they are doing well again and their business is running well."

Enzo smiles. "We're so happy we could help and so, so honored, by the way, to be seated at your table. It makes us both feel almost like… family."

"Well, you know, you're welcome and what's more… I think the kids wanted us to get to know you both better," says Alejandro.

Now, Valencia quickly wraps up a superficial conversation she is having with the person seated next to her, excusing herself for a moment, and she leans in to say, "Well, you know, the best way to do that is for your family to come visit us… say in the spring."

Richard Harthun now leans over to say, "Where? In Italy? Bring the family to Italy? Wow, that's a long trip and getting the kids together and finding hotels, etc…"

"That's not a problem," says Enzo. "We have a family villa in Tuscany that can easily accommodate a good sized group. We're happy to have you stay with us. Think…what a wonderful spring vacation for you and your family with our family. We have friends and relatives visit all the time. It would be our honor."

Richard looks at Alejandro, "Wow. How about that? That would be like a postcard vacation. Always wanted to see Tuscany." Alejandro agrees and chuckles.

"Wonderful," says Valencia. "Let me tell your ladies that it's all set, except for the dates. I think they will love it, and I can't wait to introduce Savannah to some of my Italian designer friends."

As the festivities slowly die down, many people have already gone and Hudson and Jamie are getting ready to depart for their hotel before leaving tomorrow for a honeymoon riverboat cruise through France. They are saying goodbye to guests and stop by to see Enzo and Valencia Costa. Jamie starts, "Well, we are leaving tomorrow, pretty early, for our riverboat trip through France, but we wanted to tell you how delighted we were that you could be here today."

"Oh, we wouldn't have missed this for the world, Jamie," says Valencia, who then gives Hudson and Jamie a very heavy gift envelope.

"Oh, that's really not necessary because your husband gave us terrific 'gifts' previously," says Jamie.

Valencia replies, "I know that, dear, but here's some advice from an older couple; never turn down money, even if it's it just for an American hotta dogga. We want you both to have this." Jamie and Hudson are both surprised and filled with gratitude. Valencia continues, "You two have a wonderful trip and perhaps we'll see you in the spring."

Hudson says, "The spring?"

"Yes, yes…talk to your parents. We have something planned," finishes Valencia. "Now, you go."

CHAPTER 16
Yin Yang

Ancient Chinese philosophers observed contrary or opposing positions in all things known to human beings. They termed this the Yin and the Yang. The conceptualization of yin and yang suggested that there was a positive and a negative side to everything and each side carried with it the potential for the other. They were complimentary, inter-dependent and formed a spectrum in the universe, as it was known. This theory can be demonstrated today in almost everything we do. Such is the case in the lives of Hudson Harthun and Jamie Broad-moor Harthun. They have had many successes in their young lives and also many stresses that, in fact, have appeared to be related.

Knowing each other for their whole lives, being essentially raised together, working together in a very supportive and varied manner, creating a company together, which was born out of similar interests and now being married and… wait for it… expecting their first child, would suggest that life has all been a bed of roses for the young couple. This would be, of course, an inaccurate conclusion—in that it's incomplete. Their company and their brilliance, as a couple working together, created a product which could be seen in a very positive light, but could also be seen as a sinister creation, with great threats to individuals, while offering salvation for the masses. A dilemma, to be sure, but not one caused by any direct intervention on their part.

Jamie and Hudson's parents are delighted at the news that Jamie is pregnant. Her sister is already designing baby clothes. Their "third set" of parents, Enzo and Valencia Costa, are equally thrilled, perhaps even more so. Valencia is so happy you would think it was a royal family about to have an heir to the throne. Her life has opened up again and her thoughts race over where the child will be schooled, how often she will get to see them, what the child will call her and Enzo, and, just about every other thing one can think of. This is to be, after all, a happy occasion. Something to look forward to with very positive expectations—the "yin." There is also, as many believe, the interdependent "yang," the negative side of life.

Sometime after the Norwegian oil rig attack, resulting in the death of the CEO of NOS (Norge Oil Subsidiary) as well as several others in his entourage as collateral damage, the next executive to step into the CEO position for NOS, Bjorn Halvorsen, is pondering the event. He has staff do some in-depth research about this and similar events, globally. When he looks at the list of events he is researching, he notes what he sees is a pattern. In conference calls with several other fossil fuel company execs, it was decided that they need to meet face to face to discuss the emerging pattern. Three other CEOs agree to meet; most do not. A time was set up and a place to meet was agreed upon. They will travel to Kuwait and meet at a time and destination known only to the four of them. This meeting will take place in the utmost secrecy and will be a brainstorming session, where each participant will be assigned a path to pursue to try to put the puzzle together. They all already believe that there is a pattern of assassinations occurring and the usual protocols for investigation always turn up empty, with more promises to continue "investigating." Bjorn's theory from the get-go was that these were inside jobs.

Someone in each of these oil companies had a serious vendetta against the CEOs and decided to hold a lethal intervention, just after there was a decision to pursue some sort of expansion of the company. These initiatives may have been opening new drilling fields or platforms, declining already pledged payments to further "green" initiatives for better climate action, destruction of forested areas, opening new coal mines, etc. It certainly appears to Bjorn that climate activists may be behind this and the others agree, but they have difficulty determining how this could be happening without successful police action, ever. Are the police somehow complicit in allowing this to go on? Are the perpetrators just so clever and careful, that they make no mistakes and leave no clues? They cannot understand how any organized effort, such as this, could be occurring globally, with complete immunity. What type of organization has the worldwide power, finances and political influence to accomplish this?

One member of the investigative group is assigned to find out how news of inside corporate decision making could travel so fast. Another's task is to pursue the police and military reports and investigative actions. Why have these investigations all ended up being closed, essentially becoming cold cases? A third CEO is assigned to pursue the media. What journalists are involved in reporting on these global events? Where do they get their information? Who do they speak to? Where do their initial stories lead them and are they pursuing the open questions and if not, why not? Bjorn, himself, decides to pursue the political arena. Someone in power should know why this is happening. Why are they not getting answers...or, perhaps they are, in which case, why are they not going public with this? Someone... someone, somewhere knows.

Each of the CEOs works alone and is extremely careful to ask only questions that are unlikely to arouse suspicion. They do not want any of this to get leaked to the media. They are on the right track. They know that a leak of this nature could trigger the next lethal intervention and they do not want to find themselves on the business end of that rifle.

The first CEO explores how seemingly corporate insider information could travel so fast. He surmises that people inside those organizations, who had access to high level decisions, must be reporting this to someone, but who? Why would anyone take on an intervention so risky and execute it, so to speak, so well? He considers organized crime, governments and, especially, fossil fuel competitors as well as other oil and gas companies or "green energy" companies.

The next fossil fuel executive follows up on police reports, including the reporting officer, the supervisory people, department chiefs, clerks, etc. He can find nothing out of order, other than a similar conclusion on each report that there is not enough information available to pursue a continued investigation at this time. It does, however, seem that a small amount of information that was uncovered was properly considered.

The third CEO has a more "meaty" task. He pursues a path dealing with the media. He is to investigate the investigative journalists, which is also not a small task, considering their usual extreme reluctance to discuss any of their sources, even when threatened with jail time. This executive goes outside the box, however. He partially discusses his task with a young private investigator, who will dive deep into the lives and writings of the few journalists who have written articles on the disappearance or confirmed murder of top oil and gas executives. He hires the private investigator, knowing this is not part of the agreement of the group he was part of. He figures this will somehow insulate him and he does not want to get his own hands "dirty." This CEO#2 seems to have the most success of the three. He finds that there is a journalist named Sara Meclin, who had started or perhaps completed an article for her magazine that involved a video game company named Harthun-Broadmoor Entertainment. He is able, after doing some superficial digging, to "gain the trust" of an insider journalist named Neil, who knows Sara well and tells him about the game company and that Sara had completed an article on them. He tells him he does not know where Sara went, but only that she quit her job to accept another position closer to her "family" where she would be happier. He does not know where that is, however.

CEO#2 is hooked. He pays Neil and asks him to search Sarah's files for any sign of that article she wrote; he assures him he will be "paid well" for it. Neil, of course, is a player and is drawing the CEO in to find out what he and his group already know. They work closely and eventually, Neil finds out who the other three CEOs are and when and where their next meeting will be. CEO#2, of course, has a big mouth and cannot keep a secret. He wants to appear to be the top man in the group and be the one who solves the problem. There is a concept, if you will, in business, called the Peter Principle, which states that "when moving up the ladder, individuals will rise to their own level of incompetence." This is clearly the case with CEO#2, not being as clever as he thinks he is. By simply inflating his monstrous ego, Neil easily sucks information from him, actually considering him a simpleton. Neil now even knows which hotel and which room the CEO group will meet in when they are in Kuwait. We have all seen people, in high places, rising through the ranks by exploiting others who don't have discerning capabilities of their own, while they, themselves, are just narcissists and not particularly intelligent. That is the case here. Stupidity, not ignorance...outright stupidity.

Bjorn, the CEO's group leader, follows up on political insiders, meeting with high level government people, lobbyists, staffers, etc. He does seem to glean some useful information, which is that he occasionally finds someone, who cuts off the conversation cleanly and abruptly and somewhat oddly. It is almost as if they were afraid of someone or something. Globally, there are other political figures who have disappeared, after all. Bjorn finds this odd, but tucks the information away for the time being, waiting to see what the others have found. It isn't long before he receives a call from CEO#2 saying that he believes he has found the Holy Grail and that he is going to impress them all at the next scheduled meeting. He is sharply and abruptly admonished by Bjorn for even discussing this over the phone, and he reminds him of their agreement to total secrecy. CEO#2 then blurts out more information right over the phone, to appear unfazed by any external threats, now that he has all but solved their problem. He is quickly speaking to a click and then silence. Over the next week or two, other "planning" for the meeting is going on.

Now, in Kuwait, the four CEOs all show up for their second meeting to compare notes. They each have separate but adjoining rooms. After they are all registered, it is decided that they will all go to dinner and then meet in the morning in Bjorn's room.

CEO#2 quickly interjects that he prefers to meet in his room, rather than move his materials and flow sheets somewhere else. "I have it all planned out. It should be done in my room since I am undoubtedly the one who gathered the most important information." The others stare at him, but since it doesn't seem to matter whose room it is, they all agree.

Later, at dinner, CEO#2 starts right off discussing some of his information, only to be grabbed abruptly on the arm by Bjorn, who appears shocked at the very topic even being brought up in a public place. "Listen…listen, let's just hold any 'business' discussion for tomorrow, shall we? Now, has everyone been following the PGA tournament? My son is really into golf now and is actually getting pretty good. I can see Tiger's form in him."

The others look very nervous, now, considering CEO#2's brash, conceited disregard for their and his own safety. Bjorn knowingly glances at each of the other CEOs, and he is clearly understood without saying a word. Dinner is disbanded early to try to avoid further disclosure CEO#2 may make after a few drinks. They each retire to their rooms. After about an hour, Bjorn calls CEOs 1 and 3 in their rooms. They decide quickly to meet in the morning, along with CEO#2, as planned, but they will only listen to him and not disclose any of their own information to him. They will come up with some excuses, saying they were not able to find anything conclusive. They will agree not to take any actions, just yet, regardless of whatever information CEO#2 has and to just let things simmer for a bit and go home and mull over the data. Clearly, they will be trying to diffuse CEO#2, who is clearly a hot mess, which appalled and frightened the others.

The next morning, they meet in Bjorn's room as had been planned. They all have coffee and pastry and organize their presentation. Bjorn makes sure they all speak in low voices, almost a whisper and not in keeping with CEO#2's personality at all. He is clearly chafing at the bit to impress the others with his investigative prowess. Bjorn makes it clear how their data is to be organized and presented. CEO#2 then starts in a very histrionic manner to present the information he had gathered on Harthun-Broadmoor Entertainment and the climate game, CORPUS-A.

They listen intently to the very specific data he has gathered, until Bjorn stops him and says, "This is some very detailed information. Honestly, at this stage, this is way more than was expected. Do you mind if I ask how you obtained such detailed data?"

CEO#2, so full of himself, proudly announces, "Well, we needed some specific information. More than any of us could reasonably gather alone, without arousing suspicion, so... I obtained the best private investigator I could find, told him what we needed and after some real digging, as you can see, he came through!"

A pin could be heard dropping, in the room. Bjorn speaks out loud. "You fool! Do you realize what you've done, other than being a blow-hard, I mean? You've let the cat out of the bag and now who knows how many others are privy to our investigation! This meeting is over and we will decide later how to proceed! I'm leaving today. What say you?" he addresses the others. He notices CEO#3's eyes are almost closed, as if he did not sleep well last night. "What the...? Are you okay? Is this direct threat to our safety boring you?" CEO#3's eyes are now almost closed and he is wobbling in his chair. "OMG!" says Bjorn. "Come help me with him! He's going to fall!"

Actually, he does fall right out of his chair and Bjorn notices the other two are about to do the same thing. Bjorn panics when he realizes that something sinister is happening. He runs to the door to his room to quickly escape, only to find it locked. He runs for the room phone and can't read the lettering, telling him how to reach the front desk. He just hits "0" and is talking to a dial tone. The panic is subsiding and Bjorn is very relaxed and sleepy. He falls back into a sofa, for a quick rest... his final one.

The next day, when the cleaning staff has not had access to the room for over 24 hours, they contact the hotel management, who checks the room to find all four CEOs dead. After an investigation by the authorities, it is determined that they were all poisoned by an odorless, colorless gas that probably came from a vent, but no equipment is found. Post mortem exams found the bodies to all be somewhat odd cherry colored and their blood was saturated with CO, carbon monoxide. The coroner's report is very suddenly lost only to resurface fairly quickly with a completely different conclusion, one that is much more palatable to the hotel owners and their public relations staff and one that does not reek of foul play.

This news, however, spreads quickly, with all sorts of wild, or maybe not so wild, suggestions of foul play and conspiracy theories. All the major news media outlets pick this up as a top story. The internet is abuzz. This, of course, is not lost on Neil, the fellow journalist friend of Sara Meclin and as he is reading about it, he just shrugs with a sigh of relief, knowing how and why the cause of death here was reported as it was. Also, because this is a top story, the other story, way down the list, of a local private investigator who recently tripped "accidentally" in front of a light-rail train and died, does not even make it into public consciousness. The information loop had been closed… tight.

Jamie and Hudson also take note of this event. It would be impossible not to. They have become used to these events, in the media, much as the general public becomes used to the next gang murder, the next pack of political liars or the next theft of millions of dollars by some pharmaceutical company. After a while, this all becomes white noise and Jamie and Hudson have more important things to concern themselves with, such as picking a name for their baby, where to buy a house, what the safest SUV would be for their expanding family, etc. These are the really important things in their lives right now.

In the meantime, their "game" is reformulating its targets.

ABOUT THE AUTHOR

Travers Scott Knight is a physician renaissance-type man. After many years of really listening to his patients, he has internalized many of their most pressing concerns and has been able to put them on paper to create characters that all of us can relate to and whom create an immediate emotional impact. He has traveled widely, lived and explored many parts of the country and has wide interests. He currently finds the climate problems facing the country, to be almost insurmountable … almost. Heartened by the younger generations' zeal to take charge of their own future, Knight has created a fictional work appealing to young and old alike.

Living in the west, Knight has many hobbies and takes advantage of the best that outdoor life has to offer, including skiing, mt. biking, surfing and paddleboarding and tries to incorporate all life experience into his work. Family has always created an emotional drive, for Knight and continues to figure into the design of his work.